INTERNATIONAL ENERGY AGENCY

STEAM COAL

PROSPECTS TO 2000

This Study is published on the responsibility of the Executive Director of the International Energy Agency of the OECD. The projections were made independently by the staff of the Secretariat using methods of analysis, assumptions and information of their choice. Therefore, the results and views found in this Study do not necessarily reflect those of IEA Member Governments.

ORGANISATION FOR ECONOMIC
CO-OPERATION AND DEVELOPMENT

2, rue André-Pascal - Paris

1978

The International Energy Agency (IEA) is an autonomous body which was established in November 1974 within the framework of the Organisation for Economic Co-operation and Development (OECD) to implement an International Energy Program.

It carries out a comprehensive programme of energy co-operation among nineteen* of the OECD's twenty four Member countries. The basic aims of IEA are:

  i) cooperation among IEA Participating countries to reduce excessive dependence on oil through energy conservation, development of alternative energy sources and energy research and development,

 ii) an information system on the international oil market as well as consultation with oil companies,

iii) cooperation with oil producing and other oil consuming countries with a view to developing a stable international energy trade as well as the rational management and use of world energy resources in the interest of all countries,

 iv) a plan to prepare participating countries against the risk of a major disruption of oil supplies and to share available oil in the event of an emergency.

* IEA Member Countries: Austria, Belgium, Canada, Denmark, Germany, Greece, Ireland, Italy, Japan, Luxembourg, Netherlands, New Zealand, Norway, Spain, Sweden, Switzerland, Turkey, United Kingdom, United States.

The Organisation for Economic Co-operation and Development (OECD) was set up under a Convention signed in Paris on 14th December, 1960, which provides that the OECD shall promote policies designed:

— to achieve the highest sustainable economic growth and employment and a rising standard of living in Member countries, while maintaining financial stability, and thus to contribute to the development of the world economy;

— to contribute to sound economic expansion in Member as well as non-member countries in the process of economic development;

— to contribute to the expansion of world trade on a multilateral, non-discriminatory basis in accordance with international obligations.

The Members of OECD are Australia, Austria, Belgium, Canada, Denmark, Finland, France, the Federal Republic of Germany, Greece, Iceland, Ireland, Italy, Japan, Luxembourg, the Netherlands, New Zealand, Norway, Portugal, Spain, Sweden, Switzerland, Turkey, the United Kingdom and the United States.

TABLE OF CONTENTS

LIST OF TABLES

Note : **Metallurgical Coal or Coking Coal** *is any coal used in coke ovens for the production of metallurgical coke regardless of its chemical composition.*

*   **Thermal coal** or **steam coal** is coal used for purposes other than in coke ovens regardless of its chemical composition. Thermal coal and steam coal are used interchangeably in this report, except in discussion of the coal trade for power generation where steam coal is preferred.*

*   The distinctions drawn in this report between **hard coal** and **brown coal** is explained in footnote page 33.*

A

EXECUTIVE SUMMARY

THE MAIN CONCLUSION

A massive substitution of oil by coal will be required of industrial societies and developing countries alike if they are to sustain in this century even modest economic growth in a setting of moderately increasing energy prices. This displacement of oil by coal, however, will in itself not be sufficient but must be accompanied by more rigorous energy conservation and supply development.

Without the contribution of both expanded coal usage and vigorous demand and supply adjustments, energy needs will within a decade require more oil than major oil exporters may be willing to produce. In that event, the future energy market will clear only by some combination of sharply higher prices, slower growth, higher unemployment, and a costly transformation of industries. While higher prices would permit the entry of higher-cost deposits of oil, natural gas, and — ultimately — synthetic fuels, these sources of energy are not nearly as plentiful, dispersed, nor, in the case of synthetics, technically or commercially perfected, as coal. As compared to coal, the other sources would come too little and too late to permit a smooth transition to higher energy prices in the latter part of this century.

The limit on the re-introduction of coal is certainly not its reserve base, for exploitable resources are great compared to needs to the year 2000 and beyond. Nor as serious as they are, the environmental concerns about expanded coal trade and usage should not greatly impede coal expansion. There have been significant advances in technology in the mining, shipping and burning of coal in an environmentally acceptable manner while maintaining its cost competitiveness. The enlarged coal trade foreseen in this study, however, will require for its realization further advances to minimize new stresses upon the environment.

The principal impediment appears to be a failure or slowness on the part of many private and public energy planners to appreciate the growing economic attractiveness of using coal instead of oil. The potential of coal has been radically transformed by higher energy prices; the potential will be greater still with moderately increasing prices. When coal becomes an important internationally-traded energy commodity it will be in many instances more cost-competitive than any energy source then in abundance. But the creation of an immensely expanded world trade in coal is in part dependent upon the adoption and execution of co-ordinated government policies to facilitate coal development and usage. The next chapter outlines some suggested lines of policy action for IEA Member countries.

THIS STUDY: ITS SCOPE AND PURPOSE

In early 1977 certain disturbing medium-term trends in the world energy market were gaining strength. These trends were:

- the slowing in the delivery of nuclear capacity, thereby postponing significant substitution of oil;

- the prospect that the developing countries and the centrally planned economies might also require larger than expected oil imports;
- the likelihood that the principal oil exporting countries sometime after 1985 will produce less oil than assumed earlier; and
- the probability that on a global basis potential excess demand under current trends would place growing pressure on prices after 1985.

Unfortunately, these trouble-laden trends were accompanied by an ebbing in public concern about medium-term energy problems, promoted by the short-term slackness in the world oil market arising from long-developing new supplies and lagging economic growth.

Therefore, renewed attention at the International Energy Agency was directed to coal as a conventional fuel offering a significant contribution in dislodging oil, especially in power generation and industrial process heat. Despite the current disequilibrium in the energy market that produces many uncertainties about the future course of the market, it was thought important to complete this study and stimulate new thinking now toward coal usage and development in order to allow sufficient time for reshaping energy policies and transforming them into changed behaviour in the market place. Such thinking is the first step toward avoiding any hasty resolution of the potential energy imbalance in the 1980s that would impair further the already reduced prospects for economic growth in the last two decades of this century.

The study contains projections to the year 2000 based on two cases. The first is the reference case, embracing alternative assumptions of lower and higher nuclear capacity expansion that capture the extent of uncertainty about future nuclear development. The second case — the enlarged coal case — is characterized by a more intense policy commitment to faster coal development and more extensive substitution of oil by coal.

Both cases, however, are based on identical assumptions about economic growth rates, future real energy prices (constant to 1985 and rising thereafter by 2.5 per cent yearly); conservation effort, and electricity demand growth. The details of the projections are set out in Chapter C and the results are summarized below.

THE FINDINGS: ENERGY GENERALLY

The overall results of the reference case projections show that, even with the successful implementation of nuclear power programmes implied in the high nuclear sub-case, and in spite of the more than two-fold increase in OECD coal consumption expected between 1976 and the latter part of this century, projected energy requirements are not fully met by the sum of the expected supplies of indigenous and imported fuels. The quantity of oil that oil-exporting countries may make available for export is insufficient to satisfy expected demands, and the result in both nuclear sub-cases is a potential excess demand for energy that rises steadily throughout the period after 1985.

Energy requirements in the OECD are projected to grow by 2.7 per cent per year through the year 2000, while economic activity expands at an average annual rate of 3.4 per cent over the same period. In the OECD's energy supply sectors, indigenous oil production increases until 1985, then declines. Natural gas production declines steadily throughout the period. Therefore, substitution of other forms of energy becomes increasingly necessary. Nuclear energy grows at average annual rates of 7.9 per cent and 9.0 per cent in the low nuclear sub-case and the high nuclear sub-case, respectively, after 1985. Coal demand grows 3.2 and 2.7 per cent annually in the low and high nuclear sub-cases, respectively, from 1976 through 2000.

In the reference case, between 1976 and 2000, OECD coal utilization in the low nuclear sub-case rises by 1096 Mtce (767 Mtoe), the equivalent of almost 15 million barrels of oil per day (Mb/d). As Table A-1 shows, substantial increases in thermal coal utilization occur in every region of the OECD and in several industrial sectors as well as in power generation. The more than doubling of thermal coal use within 24 years foreseen in the reference case is, itself, an ambitious task that will require substantial new investments in coal production, transportation, and utilization projects.

Table A-1 **OECD Thermal Coal Utilization**
Reference Case (Low Nuclear)

OECD Region	Total Uses			Power Generation Only		
	1976 Mtoe	2000 Mtoe	Per cent change	1976 Mtoe	2000 Mtoe	Per cent change
North America	308	705	129	269	592	120
Europe	159	371	133	126	297	136
Pacific	32	116	263	30	118	293
OECD Total	498	1 193	140	425	1 007	137
IEA Total	455	1 087	139	391	922	136

When the demand and supply components of the reference case projections, including the very large expansion of coal utilization, are aggregated to derive the OECD's potential net oil import requirements, the low nuclear sub-case produces a net oil import demand of nearly 33.2 Mb/d in 1985, increasing to 40 Mb/d in 1990 and 49 Mb/d in 2000. After adding estimates of the net oil import demands of non-OECD regions, total potential world demand for OPEC oil would amount to 35 Mb/d in 1985, 42 Mb/d in 1990, and a wholly unrealistic 56 Mb/d in 2000. The successful implementation of nuclear power programmes implied in the high nuclear sub-case would only reduce potential world demand for OPEC oil by 1Mb/d in 1985 and 1990, and 2 Mb/d in 2000. Even with the more favourable assumption about the pace of nuclear expansion, potential world excess demand still comes to nearly 8 Mb/d in 1990 and 24 Mb/d in 2000.

The reference case shows that — in spite of a very significant increase in coal utilization — substantial potential excess energy demands exist in both nuclear sub-cases. An enlarged coal case, with even more extensive substitution of coal for oil, was constructed to show how much of the potential excess energy demands of the reference case could be met by the adoption of more intensive coal policies and programmes favouring an even faster increase in coal utilization.

Enlarged coal case

If industrial countries implemented coal substitution policies in the years immediately ahead such as those recommended in Chapter B, a massive substitution of coal for oil could be realized toward the end of this century. With due consideration for the technical and economic limits to coal substitution determined by process requirements, lead times, environmental restrictions, and the oil price assumption of this study, realization of the substitution potential shown in the enlarged coal case would increase OECD thermal coal utilization over the reference case (low nuclear sub-case) levels by 100 Mtce (70 Mtoe), in 1985, 219 Mtce (153 Mtoe) in 1990, and 504 Mtce (353 Mtoe) in 2000.

The additional coal utilization shown in the enlarged coal case would permit corresponding reductions in demands for oil and natural gas. If the freed-up natural gas were also substituted for oil, OECD oil demand and net oil import requirements could be reduced by 1.4 Mb/d in 1985, 3.0 Mb/d in 1990, and 7.0 Mb/d in the year 2000. Even without any additional policy measures to encourage conservation or the expansion of other energy supplies, the enlarged coal case would reduce the potential imbalance between world oil demand and supply by 35 per cent in 1990 and 28 per cent in 2000.

During the period in question, coal cannot satisfy the entire potential excess demand shown in the reference case. The remaining potential imbalance between energy supply and demand not filled by coal will require more vigorous policies to promote additional conservation and expansion of other energy supplies, but the burden placed on these alternatives can be alleviated considerably by concerted efforts to promote the expansion of coal utilization and trade.

Thus, coal emerges as an attractive substitute that has the potential, with the proper policy support, to make a significant contribution by 2000. Furthermore, increased utilization of coal in the near future may speed the development of new technologies that would allow even greater substitution oil by coal-based liquid and gas fuels for oil in the latter part of this century.

OTHER FINDINGS

Perception of the Future Coal Market

The future worldwide coal market remains indeterminate at this time to most potential consumers and producers of coal. Until there is a clearer perception of the potential world coal market, and of supporting government policies and programmes, consumers and producers will not make commitments in the coal trade of the order of magnitudes foreseen in the enlarged coal case. Consumers and producers will have to be convinced that public policies will clearly support expansion of coal production, increased coal utilization and development of a transportation infrastructure for coal, in order to initiate full-scale development programmes for expansion of coal trade. Due to the lead times required for expanded coal development and utilization, public policies must be clearly defined now to impact on the world energy situation in 1990.

Principal Policy Areas

Shifts to coal by industry require considerable lead times, particularly in the electricity generation sector, and fuel choices made in 1985 and 1990 will generally predetermine fuel shares to the year 2000. Therefore, public policies must be formulated now to achieve a favourable balance in fuel shares for the beginning of the 21st century. The existing constraints and corrective policies are spelled out in Chapter B. The principal policy areas are: development and trade; material and fuel substitution; conversion to new technologies; and the environmental considerations related to all of the foregoing. Only by developing clear, co-ordinated policies in all of these areas will it be possible to achieve a programme which will utilize world coal resources to their fullest advantage in an environmentally acceptable and economic manner.

Development and Trade

Faster development of mining must await development of demand but once demand begins to grow strongly, present leasing schedules, resolution of environmental problems, fiscal regimes and initially rail and slurry pipeline infrastructure could become constraints upon expansion unless corrective policies are implemented. As international trade grows, port and inland waterways will need expansion. But the precedent expansion of demand too requires a boost for incentives for the needed capital investment, amended ordinances to allow creation of cogeneration and energy parks, and above all workable environmental rules on fuel choice, emissions control, and land usage. An agreement among IEA Member governments to ensure freer trade in coal would assist in promoting confidence in traders and investors in an unobstructed international trade.

Fuel Substitution

The greatest potential for fuel substitution by coal lies with large fuel users because of logistical and environmental considerations. The long-term potential for coal substitution will be greater when capital stock is replaced in a period of sustained economic growth. The most apparent near-term potential for coal lies in the electricity generation sector. Comparative estimates of electricity generating costs have been developed for new oil, coal and nuclear plants in order to determine the competitiveness of coal; in all cases coal is more cost-effective than oil for power generation, and in many instances, coal is more cost-attrative than nuclear. It is considered that coal will be competitive even at prices considerably above present levels. Coal stations require less construction time and less capital than nuclear stations with less long-term commitment for controversial issues of nuclear power to be resolved. Also, substitution opportunities for coal burning to generate process heat for industry, or to substitute coal-fired electricity for oil and gas boilers in industry, also appear encouraging from available data.

New Technologies

Some of the new technologies presently under development promise significant potential for conversion of coal and new or alternative energy sources by 2000, or earlier if progress is made in reducing their supply costs. Such new technologies relate to coal mining, fluidized bed combustion of coal, environmental controls, production of low-BTU gas from coal, fuel cells utilizing coal-derived fuels and improved coal transportation. All of these technologies must be considered individually as they will replace existing technology wherever they can deliver energy to a market at a competitive price. This report assesses the time required for significant commercial impact on energy production and the degree of market penetration expected before 2000. Cost estimates have been included where possible.

Environment

Environmental problems stemming from coal mining, transportation and utilization and particularly degradation of land, air and water as well as increased water consumption have been addressed in the light of what has been done, what is being done and the costs involved.

The known environmental impacts of the entire coal cycle may be divided into the health and safety risks (those of combustion to the general public and of mining accidents to miners) and the physical risks (those of damage to land, water, plants and the ecosystem). The second class of risks are better understood. For most of these, manageable control technology exists although institutional arrangements for assessing and managing these risks are often inefficient and needlessly time-consuming. The first class of risks, the health and safety risks, may not be as amenable to further substantial reduction by existing technology. Furthermore, other possible risks in the coal cycle, of unknown character or magnitude, are being researched, such as those linked to carbon dioxide, heavy metals and synergistic effects, but as yet evidence is not such as to offer any policy guidance for coal development. For example, a public concern has risen over trace elements in coal such as lead, cadmium and uranium; however, it is not yet established whether these elements are released into the atmosphere during combustion or retained in the ash.

Overall the expanded use of coal will require some compromise or harmonization between environment and energy objectives, with careful selection of control measures that are most cost-effective, and with the higher costs borne by energy consumers. The costs of meeting current environmental standards are given in this study and are taken into account in the comparative costing of utility fuels.

B

LINES OF POLICY ACTION

The first result of this study of the potential for substitution of coal for oil has been the identification of constraints on coal development — some are already acting to slow development of coal usage and supply and others are anticipated to do so as trade rises to higher levels. While many constraints are of a technical nature having to do with such matters as inadequate infrastructures or cumbersome institutional settings for obtaining permits, the initial, pervasive constraints have to do with perception of the future energy market and public attitudes toward coal, combined with inertia and failure to re-examine choices despite a radical alteration in energy needs, available supplies, and costs. And the second result is a search for policies which might remedy these constraints.

In today's setting of an apparently abundant oil supply, a consequence of long-developing new sources of oil from Alaska, the North Sea and Mexico, matched by slow economic recovery, there is a general reluctance to face costly choices to solve supply problems that may emerge only ten years later. Behind the present reluctance to invest more in coal production, there is some public disbelief that oil supply in the future will be insufficient and therefore more costly, or if more costly, that all energy prices will advance together leaving little preference among fuels on cost grounds. While such attitudes are probably most prevalent among the public, they are occasionally held by utility and industry decision makers and have led to postponed orders for generating capacity of both nuclear and coal, and have, to a lesser extent in the utilities than in general industry, led to a reluctance to switch from oil or gas to coal. Increased production and trade will only be possible if there is sufficient demand. Therefore, stimulation of demand should be one of the main objectives of government policies.

The second area of attitude that inhibits coal development is inherent in the bulkiness, dirt, and pollutions long associated with coal. This attitude has been strengthened by heightened concern about environmental risks associated with combustion of coal. In response to this concern much environmental legislation was enacted, some before the higher energy prices of 1973/74, but experience has shown that while the objective of restoring and maintaining a safer, less physically disturbed environment was an undisputably desirable objective, the control measures chosen to achieve the objective were sometimes not cost-effective when enacted. And little of the legislation or indeed of public attitudes has been re-examined since as to the socially desirable balance between environmental and energy objectives. The results of this study should not be interpreted as an argument for the lowering of environmental standards or slackening in their enforcement, but instead for cost-effective enforcement with some account taken of the impact upon energy objectives as well as environmental objectives.

This study goes to some lengths to portray the consequences of energy trends in a reference case based on assumptions uniquely favourable to lower energy needs and higher supply — in other words, favourable to the minimum potential imbalance of demand and supply. It concludes that a much expanded coal trade is not only necessary to resolve such a potential imbalance, but in itself not sufficient and must be coupled with renewed efforts in greater conservation than the already greatly reduced demand assumed here, and more other conventional and non-conventional energy supplies in order to sustain economic growth in both industrial societies and developing countries along a transition path of moderately increasing energy prices. The economies of coal compared to other energy forms appear so favourable as to ensure realization of at least the trade seen in the reference case; but new policies are needed to promote the needed international trade, regulatory, price and tax climate to foster even larger, and earlier development of coal trade.

While government policy actions in developed countries to support increased coal utilization and trade include a wide range of options, they might nevertheless be categorized as follows:

1. Policies that encourage substitution of coal for oil in power generation, industrial processes, and space heating. These might include:
    —   prohibition (whenever economically and engineeringly feasible and with a minimum of exemptions) of the construction of new or replacement facilities for oil or gas-fired baseload electricity generation and in other major fuel-burning installations; conversion, if economically and engineeringly feasible, of existing oil-fired capacity and its progressive limitation to middle or peak load; and financial incentives to upgrade existing coal-fired units and accelerate conversions to coal;
    —   support for research, development, and commercilization of promising emission control technologies, coal cleaning and blending techniques, fluidized-bed combustion, ash control and disposal systems, and other technologies that might improve the practical aspects of coal combustion and allow greater coal substitution in a wider variety of uses;
    —   improved co-ordination between energy and regional and urban development planners to promote coal-based energy systems to serve the needs of new industrial parks and large residential, commercial and government building complexes through district heating and co-generation schemes.

2. Policies that promote a more favourable climate for investment and encourage the timely development of coal production. These might include:
    —   clarification of environmental regulations; development of more expeditious procedures for settlement of environmental concerns; acceleration of leasing schedules under conditions ensuring sufficient land restoration and water protection; speedier resolution of native claims;
    —   co-ordination and encouragement of the planning and development of coal transport infrastructure (rail, pipelines, inland waterways, ocean carriers, loading and receiving facilities);
    —   financial support of coal development projects comparable to the assistance provided to other energy supplies;
    —   moderation in fiscal regimes (including government royalties, severance taxes and railway tariffs) relating to coal development.

3. Policies that clarify existing uncertainties and promote greater confidence among investors and traders to encourage an expansion of international trade in steam coal. These might include reliable arrangements, to ensure and respect the free movement of steam coal, and investment in coal projects, across international borders, taking into account policies to maintain domestic production for reasons of security of supply, social and regional concerns but in ways which permit that part of demand that would not be covered by the desired indigenous supply be met by imported coal rather than oil imports.

4.   Policy actions to implement effective information programmes, aimed at both the general public and investors and traders, to increase public awareness of energy policy issues and the trade-offs between national security and relative economic costs, including environmental health and safety protection costs, associated with coal and other fuels.

Some of these types of policy action will apply to all regions of OECD, whereas others are more relevant to specific countries only, or apply with different intensity and sometimes in different time periods in various OECD countries. These distinctions are made in subsequent sections on individual Member countries or technical subjects.

With respect to developing countries, depending on their interest in expanding coal use and trade, policy actions might include the following:

—   creation of a better understanding of the coal cost-benefits and favourable spill-over effects from developing indigenous coal resources: e.g. employment for miners, creation of rail-road lines, and saving of foreign exchange reserves by displacing oil imports.

—   multinational arrangements, in conjunction with international development banks and regional energy commissions, to facilitate technical and economic support for coal development projects.

—   international research and development to share information and devise new methods for overcoming the technical problems of coal extraction and use.

C

COAL IN FUTURE ENERGY MARKETS

FUTURE ENERGY MARKETS

Future prospects for coal can be properly assessed only against the background of overall world energy demands and supplies. In order to provide the total energy context, a reference case of energy demand and supply was constructed for each OECD country for 1985, 1990 and 2000, reflecting the very considerable increase in coal usage foreseen under existing policies and market trends. Aggregated projections were also made for the rest of the world[1]. Within this case, separate sub-cases were developed based on lower and higher estimates for nuclear power supply in OECD countries. There is in addition an enlarged coal case that was constructed to suggest the extent to which coal can substitute for oil as a result of even further government encouragement.

Economic Growth Assumptions

For both cases OECD GDP is assumed to grow at 3.9 per cent per year from 1976 to 1985, 3.5 per cent from 1985 to 1990 and 3.0 per cent from 1990 to the year 2000.

In the absence of official OECD long-term economic growth forecasts, these projections were constructed by the Combined Energy Staff solely for use in the present report, using the following approach. First, for each of the seven largest OECD economies projections were made for labour force growth to the year 2000. Second, for the same countries projections were made for labour productivity. The growth of productivity for each country was assumed to be constant from 1976 to 2000 but at a slower rate than over the period 1968 to 1973. The choice of slower rates reflects a growing belief that productivity growth will be reduced by environmental controls, the increases in energy prices which have occurred since 1973, and by other considerations such as desires for more leisure time. Third, growth rates for GDP for the period 1976 to 1985 were selected taking account of currently unutilized capacity and labour and of continuing inflation and balance of payments problems. Given the assumptions about labour force and slower productivity growth these rates would result in reduction of, but not complete elimination of, labour market slack by 1985. After 1985 GDPs for the main countries were assumed to grow at rates equal to the sum of projected labour force growth and assumed productivity increase.

---

1. The energy projections in this study were prepared for purposes of the coal study alone. They are not necessarily those that will appear in the revised **World Energy Outlook** *to be completed in late 1979. The latter may employ different assumptions about economic growth, conservation and supply as a result of ongoing analysis. These amendments will produce a balance of demand and supply in global energy trade in the next* **Outlook.** *In this study only the enlarged coal use case is considered in reducing the potential excess demands shown in the reference case.*

Table C-1 **GDP Growth Rates**[1]

Per cent Per Year

	1976-1985	1985-1990	1990-2000
Canada	4.1	3.4	3.1
United States	3.7	2.9	2.2
Japan	5.9	5.0	4.0
Australia	4.0	4.0	3.6
New Zealand	3.1	3.0	2.7
Austria	3.5	3.5	3.2
Belgium	3.7	4.0	3.6
Denmark	3.7	4.0	2.7
Finland	3.2	3.0	2.7
France	4.0	4.1	4.1
Germany	3.7	3.7	2.9
Greece	5.6	5.3	4.8
Iceland	4.0	5.0	4.5
Ireland	3.5	4.0	3.6
Italy	3.6	3.8	3.5
Luxembourg	3.9	3.5	2.7
Netherlands	3.3	2.8	2.5
Norway	4.4	4.1	3.7
Portugal	4.0	5.0	4.5
Spain	3.6	4.0	3.6
Sweden	3.0	2.7	2.4
Switzerland	2.3	2.8	2.5
Turkey	7.0	7.0	6.3
United Kingdom	2.7	2.1	2.2
OECD Europe Total	3.6	3.7	3.4
OECD Total	3.9	3.5	3.0
IEA Total	3.9	3.4	2.8

1. The assumed growth for the IEA for the period 1976-1990 is lower than the rate underlying the energy forecasts submitted by member governments for the 1977 IEA reviews of national programmes. The total IEA growth assumed in the submissions is 4.3 per cent per year from 1976 to 1985 and 3.5 per cent per year from 1985 to 1990.

These procedures resulted in economic growth rates for the main countries that are generally lower than those underlying the energy forecasts submitted to the IEA for the 1977 reviews of national energy programmes. For the remaining OECD countries the rates were adjusted downwards by the same proportionate amount as the average for the larger countries.

Oil Prices

International oil prices are assumed to remain constant in real terms until 1985 at the 1977 level represented by $ 12.70 for Arabian API 34° crude oil. Thereafter they are assumed to rise at about 2½ per cent per year reaching $ 18.40 (in 1977 dollars) by 2000. Natural gas prices are assumed to follow oil prices closely. But coal prices are assumed to grow less rapidly since supply curves are believed to be relatively flat in those OECD countries which have the most potential for expanded production (United States, Australia and Canada), and competition is thought to be sufficient to keep prices approximately in line with costs. Costs of nuclear generated electricity are also assumed to grow less rapidly than oil prices; these costs are discussed in Chapter F.

Energy Policies

The reference case is designed to trace the consequences of continuation of existing energy policies and market trends. An exception was made to display the results of nuclear energy capacity being delivered at current and faster rates. The reference case is not meant to be the " most likely" case, and certainly not the preferred case. It shows the growth of OECD demands and supplies if new policies are not adopted and if global oil supplies do not expand as rapidly as potential demand.

The enlarged coal case displays the opportunities for coal expansion if the general lines of policy action outlined in Chapter B were adopted by member governments.

Methodology

The reference case projections for IEA countries for 1985 and 1990 are based to a large extent on the energy balances provided by IEA governments for the reviews of national programmes. But several modifications were made. First, energy and electricity demand estimates were adjusted downwards to correspond to the Secretariat's lower economic growth rate assumptions. This was done by reducing energy demand in each final use sector by the change in GDP multiplied by the energy/GDP elasticity for that sector. The reduction in energy demand was assumed to take the form of a fall in oil demand. The second modification consisted of substituting nuclear power projections developed by the Secretariat in the lower and higher nuclear sub-cases in place of the projections contained in the submissions. The Secretariat's projections are generally lower, even in the higher nuclear sub-case. This modification usually required offsetting changes in amounts of other fuels used to generate electricity (except in fortuitous instances where reduction in total electricity demand attributable to reduced GDP turned out to be equal to the reductions in nuclear produced electricity). For 1985 and 1990 those changes were almost entirely confined to oil consumptions. The third modification involved the separation of coal demand and supply estimates into metallurgical and thermal categories for countries which did not provide a breakdown. Also in a few instances the Secretariat altered the metallurgical coal projections on the basis of information obtained after the submissions were received.

Reference case projections for 1985 and 1990 for non-IEA countries, and projections for the year 2000 for all countries, were prepared by the Secretariat. Since 2000 is a long time away, and since governments did not provide their own estimates for that year, the projections given for the year 2000 are subject to a much greater degree of uncertainty than those for earlier years.

Overall Results

The overall results for the reference case are displayed in the summary Table C-2 which reports projections of supply and demand for 1985, 1990 and 2000. Ordinarily, energy projections of demand and indigenous supply would be balanced by net imports thus achieving a "closing" of supply and demand at an assumed price. This table is a departure from that convention in the sense that supply and demand are not balanced. Imports are insufficient to achieve balance because it is assumed that oil exporting countries of other regions will not make available supplies of crude oil sufficient to satisfy potential world demand for their oil. This unsatisfied demand is reported in the last line of the table as the potential world excess demand.

On the basis of the assumptions and projection methodology total energy demand in the reference case is projected to grow at an average of 2.7 per cent per year from 1976 to 2000 compared with an average annual economic growth of 3.4 per cent

Table C-2   OECD Energy Demand and Supply and World Oil Trade

Reference Case : Alternative Nuclear Sub-Cases

Mtoe

	1976	1985	1990	2000
OECD Demand				
Solid Fuel (low nuclear - high nuclear)	708	993	1 199-1 160	1 472-1 313
Oil and NGL (low nuclear[1] - high nuclear)	1 909	2 486-2 458	2 756-2 702	3 154-3 079
Natural Gas	681	798-797	852-850	865-857
Nuclear (low-high)	88	295-324	447-542	925-1 172
Hydro/Geothermal (low nuclear - high nuclear)	226	287	324	361-356
Others		5	14	136
Total Demand	3 613	4 864	5 592	6 913
OECD Indigenous Supply				
Solid fuel (low nuclear - high nuclear)	687	952	1 130-1 091	1 365-1 233
Oil and NGL	597	824	777	699
Natural Gas	663	659	645	549
Nuclear (low - high)	88	295-324	447-542	925-1 172
Hydro/Geothermal (low nuclear - high nuclear)	226	287	324	361-356
Others		5	14	136
Total Indigenous Supply (low nuclear - nigh nuclear)	2 262	3 022-3 051	3 337-3 393	4 035-4 145
OECD Net Imports				
Solid fuel (low nuclear - high nuclear)	31	41	69	107-80
Oil and NGL[1] [2]	1 322	1 566	1 979-1 925	2 455-2 380
Natural Gas	24	139-138	207-205	316-308
Total Net Imports (low nuclear - high nuclear)	1 377	1 746-1 745	2 255-2 199	2 878-2 768
Net Oil Imports (Exports) by World Region				
OECD[2]		1 566	1 979-1 925	2 455-2 380
Centrally Planned Economies		—	—	—
OPEC[3]		(1 715)	(1 685)	(1 545)
Non-OPEC Developing Countries and Others		106	136	350
World Potential Excess Demand (Supply)		(43)	430-376	1 260-1 185

1. Includes bunkers.
2. For 1985 assumes IEA Group Target of 26 Mb/d (excluding bunkers) achieved through intensified conservation and supply expansion programmes. Without the constraint imposed by the target the balance would yield net oil import demands of 33.2 Mb/d and 32.7 Mb/d for the lower and higher nuclear sub-cases respectively.
3. OPEC production is assumed to be 38.5 Mb/d in 1985, 37.9 Mb/d in 1990 and 40 Mb/d in 2000. These assumptions were adopted after an analysis of each country's reserve base, existing and planned capacity, economic development, and oil revenue needs. Much doubt about the economic bounds on optional production was not resolved, and at least a ± 10 per cent band of uncertainty exists around these estimates. Domestic demand plus bunkers sold by OPEC countries are projected to be 4.2 Mb/d in 1985, 5.1 Mb/d in 1990 and 9.1 Mb/d in 2000.

(implying an energy/GDP elasticity of 0.79). The energy demand growth, however, decelerates considerably from 3.4 per cent per year between 1976 and 1985 to 2.1 per cent annually between 1990 and 2000. This is due to a combination of lower economic growth and increasing energy conservation.

The contribution of energy conservation is shown more explicitly for total OECD and for the seven largest countries in the tabulation below, in which energy/GDP elasticities are shown to fall progressively over time. The relatively high elasticities for the period 1976 to 1985 are partly due to the choice of 1976 as a starting point. If, for example, 1974 were chosen instead, the large energy savings achieved during 1975 and 1976 would have the effect of reducing the elasticity for total OECD to 0.79.

Energy/GDP Elasticities

	1976-1985	1985-1990	1990-2000
OECD Total	0.86	0.83	0.70
Seven Largest Countries	0.83	0.77	0.63

Changes over time in indigenous supply show some clear patterns. Indigenous oil supply increases until 1985 but declines thereafter. Indigenous natural gas supply declines steadily from 1976 on. The other forms of energy show substantial increases. Nuclear power grows at an average 10.3 per cent per year from 1976 to 2000 in the low nuclear sub-case and 11.4 per cent per year in the high nuclear sub-case. By the year 2000 solid fuel's share of total indigenous energy supply reaches 34 per cent in the low nuclear sub-case with the next largest share being that of nuclear, which reaches 23 per cent. In the high nuclear sub-case, the share of solid fuel is 30 per cent, compared with nuclear's share of 28 per cent. Hydro supply increases, and new forms of energy grow rapidly after 1990 but account for only a small proportion of total indigenous supply in 2000.

The bottom line of Table C-2 shows that net oil import demands by OECD, when combined with projected demands and assumed supplies by the rest of the world, results in a potential excess demand for oil in 1990, and 2000. As stated earlier Table C-2 depicts a partial projection; in reality no gap between supply and demand would occur in the world energy market. The potential excess demand is meant to display the magnitude of unsatisfied energy demand that must be met by an extraordinary effort in reducing demand and expanding supply.

THE ROLE OF COAL

There are several available means by which the large differences between energy demands at the end of the 20th century as projected in the reference case and the energy supplies which now seem likely to be available in the future can be reconciled:

a) *reduce* energy demands by some combination of
   - vigorous conservation
   - slower economic growth
   - market reaction to sharply higher energy prices.
b) *increase* energy supplies available by some combination of
   - expansion of the use of coal
   - expansion of nuclear power
   - expansion of hydro power
   - increased supplies of oil and natural gas
   - increased supplies of non-conventional energy sources (although they cannot be expected to make a large contribution by 2000).

Given the social and political disadvantages of slower economic growth and/or higher energy prices, the uncertainties which surround nuclear power development (not to mention its outside limitation to the electricity generation sector), the limited number of available sites for hydro power, and the uncertainties about additional oil and natural gas resources, coal emerges as an attractive alternative during the period in question.

The purpose of this study, therefore, is to examine the potential for coal as an energy source, to identify factors which constrain greatly expanded use of coal, and to suggest policies which would encourage that expansion.

Prominence is given in this report to the possible contribution of an enlarged coal trade. But a great deal more coal usage, while necessary, is in itself not sufficient to achieve this balance of supply and demand in OECD. In addition to the enlarged coal case there is need for more vigorous conservation measures, more active exploration and development of oil and gas, expansion of nuclear power, and greater effort towards accelerating commercialization of so-called synthetic fuels. Only with contributions from each of the supply and demand sectors would it be possible to balance OECD demand and supply at the assumed economic growth rates and at the assumed energy prices.

THE COAL PROJECTIONS IN DETAIL

The demand for coal can be separated into various categories including: thermal power station consumption, demand for steam coal in transportation, the residential/commercial sector, industry, coal use for transformation into synthetic gas or liquids, energy sector consumption, non-energy use of coal and finally metallurgical coal use by the iron and steel industry. These are discussed in succession below.

Electricity Generation

In order to examine the demand for coal in electricity generation, two alternative sub-cases were constructed within the reference case, one sub-case containing relatively low projections for nuclear power, and the other containing relatively high projections for nuclear power, but both assuming the same level of electricity demand and supply.

These sub-cases were developed not in order to consider coal as an alternative to nuclear power, but rather to take account of the uncertainties which surround future nuclear power growth. For the purpose of closing the gap between world demand and supply of oil both nuclear power and coal will have to make large contributions. Furthermore the larger is nuclear power's contribution to electricity generation the greater will be the amount of low cost coal available for use in other sectors.

The projections broken down by country are shown in Table 1-2 of Annex 1. Figure C-1 below displays the results for total OECD for the low-nuclear sub-case.

Total electricity supply and demand for both sub-cases are projected to grow by 4.6 per cent p. a. from 1976 (4 757 TWh) to 1985 (7 153 TWh), 3.9 per cent p. a. from 1985 to 1990 (8 669 TWh) and 2.9 per cent p. a. from 1990 to 2000 (11 518 TWh). As is the case for growth in total energy demand, the slowdown is due to projected reductions in economic growth rates and increases in energy savings.

Nuclear's share of the energy used to produce electricity is projected to grow from 8.0 per cent in 1976 (126 Mtce, 88 Mtoe) to 33.9 per cent in the year 2000 in the low nuclear sub-case (1 322 Mtce, 925 Mtoe). As a result, the share of each of the other forms of energy, including coal diminishes. Nevertheless, the absolute amount of coal[2] demanded is projected to expand from 607 Mtce (425 Mtoe) in 1976 to 1 438 Mtce (1 007 Mtoe) in the year 2000, and the absolute amount of oil and gas demanded increases from 357 Mtoe to 381 Mtoe.

Figure C-1   **Fuel Used for Electricity Generation in the OECD**

(Low Nuclear Sub-Case)

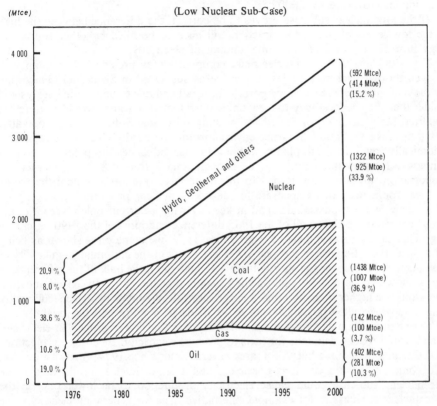

The quantity of coal increases between 1976 and 2000 but the share of coal decreases.

Table C-3   **Nuclear Capacity in the OECD to 2000**
Gigawatts at end of year

	1976	1985		1990		2000	
		Low	High	Low	High	Low	High
North America	42	112	123	158	192	301	356
OECD Europe	24	84	96	123	155	268	365
OECD Pacific	7	18	20	35	40	84	106
OECD Total	73	214	238	316	388	654	828
IEA Total	70	180	204	267	336	559	718
Difference between high and low nuclear for OECD							
(Mtce = 7 × 10¹² kcal)		42		137		353	
(Mtoe = 10¹³ kcal)		29		96		247	

2.   Including metallurgical coal products used in power generation.

The nuclear projections are displayed in Table C-3. The figures for both cases are relatively low compared with forecasts made in the past (e.g. by WAES). This is true especially for the year 2000 estimates because the relatively low nuclear capacities brought into operation in earlier years, even in the higher nuclear sub-case, would require the construction of non-nuclear generating capacity which it would not be economical to retire by 2000.

If the low nuclear sub-case proves to be correct, the difference between the high and the low levels of nuclear production will have to be filled by other sources of energy in order to produce the same amount of electricity.

For 1985 and 1990 some electricity companies may prefer to fill the gap with oil[3], even though coal may be less expensive as suggested in Section G. They may view the nuclear difference as a temporary short-fall and due to unintended delays and consider that the cheapest way to meet unintended temporary gaps would be to resort to oil-fired plant normally intended for peaking purposes. Some densely populated areas may have exceptionally strict environmental standards which cannot be met economically by coal-fired plants without further advances in pollution control technology. Finally, there may not be enough time before 1985 to construct the transportation infrastructure needed to move the large quantities of coal that some countries would need to compensate for delays in nuclear programmes.

In view of these possibilities and in keeping with a current policy scenario, the low nuclear sub-case shows all of the 1985 difference and most of the 1990 difference being taken up by oil (see Table 1-2 in Annex 1). While some oil is shown as being used to meet the difference in the year 2000 as a result of the continuing availability of the oil plant built to cover the shortfalls in 1985 and 1990, the bulk of the difference in 2000 is shown to be met by coal on the grounds that the gap between oil and coal prices would be higher then, and that nuclear shortfalls would be perceived sufficiently far in advance to allow arrangements to be made to acquire and use coal.

An important qualification to this discussion is that lower growth in electricity demand would reduce demands for fuels to produce electricity. Electricity projections for OECD countries have been decreased several times since 1974; further reduction would come as no surprise. It is possible that the lower nuclear production of the lower nuclear sub-case could cause electricity production to be lower than in the higher nuclear sub-case if, for example, alternative fuels for generating electricity are more expensive than nuclear thereby causing electricity prices to go up which would in turn reduce the quantity demanded.

Demands for Thermal Coal Outside the Electricity Sector

General

Demands for thermal coal outside the electricity sector include demands by the residential/commercial sector and transportation, industry, the energy sector, transformation into synthetic gas or liquids, and non-energy uses.

Projections for the aggregates of these demands (Table C-4) show increasing trends in the United States and Europe. These increases involve a turn around from a previous downward trend in coal consumption in the transportation,

---

3.   There is some evidence to support this suggestion. Official energy balances of IEA countries for 1985 given to the Secretariat in 1977 contain estimates of nuclear power supply which are 104 Mtoe lower than the projections given in 1976. Additional oil demand makes up half of the difference (e.g. Japan 22.2 Mtoe; Italy 14.5 Mtoe). Additional coal makes up only 6.6 Mtoe of the total difference. The rest is accounted for by reduced electricity growth and improvement in electricity conversion efficiencies.

residential/commercial and industrial sectors resulting from technical changes and competition from oil. The projected growth is due to the increase in oil and natural gas prices relative to coal prices which has occurred since 1973. Similar growth might be expected to occur in the other OECD regions, but in the absence of detailed information about fuel substitution possibilities in those regions, the Secretariat accepted the official projections (which go to 1990 for Canada and Japan and to 1987 for Australia) and extended them to 2000 without significant modification.

Table C-4 **Demand for Thermal Coal Outside the Electricity Sector**
Reference Case
Mtce

	1976	1985	1990	2000
Canada	2	1	1	1
United States	54	66	96	161
OECD Europe	52	74	88	112
Japan	7	9	10	10
Australia/New Zealand	5	4	5	7
OECD Total	120	154	200	290
IEA Total	106	137	177	260

Residential/commercial and Transportation Sectors

The use of coal in the residential/commercial sector and transportation was only 3 per cent of total energy consumption of those sectors in 1976 and will probably continue to shrink (except perhaps for large commercial establishments and where district heating is feasible) largely owing to its inconvenience for small consumers, and its disadvantages in locomotives relative to diesel fuel or electricity. However, Turkey projects substantial growth of demand for coal in the residential/commercial sector as demand falls off for non-commercial fuels.

Industrial Sector

There is, however, considerable potential for coal in the industrial sector, for example in the cement industry. In 1976 the industrial market accounted for 25 per cent of Total Energy Requirements in the OECD area and the share of coal (thermal plus metallurgical) in that market was 20 per cent. This implies that there is much room for coal expansion. It would probably not take place, however, unless coal were competitive with other fuels. In Section F it is suggested that coal is competitive with oil in thermal power plants. Consequently it should also be competitive in industrial establishments having similar scale. In addition, once fluidized bed combustion becomes commercialized (see Section G,) the attractiveness of coal for industrial undertakings of a small scale will be considerably enhanced.

On the basis of existing technology and present estimates of costs, production of synthetic liquids or gas would probably not occur on a large scale unless energy prices rise to levels higher than those assumed in the present report. There could be some " low BTU " gas produced, but its use would probably be confined for economic reasons to thermal power stations located at the gas production site. Should coal liquefaction or " high BTU " gas become competitive the demand for coal would be expanded considerably since its products could be used wherever oil and natural gas are presently being consumed. Some promising new technologies are now under study.

Energy Sector and Non-Energy Uses

There is a potential for use of coal in producing heavy oils in Canada (one estimate suggests 25 million tons might be used in the year 2000). Coal may also be used for production of hot carbon dioxide for enhanced recovery of conventional oil, and for production of heat and hydrogen to produce tar sands oil.

Coal may also be used to produce chemicals.

Metallurgical Coal

The separation of coal into two categories, metallurgical and thermal, is beset by a number of problems, not the least of which is that the OECD's *Energy Statistics* do not show such a separation in the historical data. Some of the projected energy balances obtained from IEA Governments provided estimates for metallurgical coal. For other countries, the Secretariat made its own estimates on the basis of outside forecasts and by extrapolating past trends.

Table C-5 shows metallurgical coal demand in the OECD to grow by 1.3 per cent per year from 1976 to the year 2000. These are Secretariat estimates; there are few if any official estimates of metallurgical coal demand beyond 1985. The estimates in this table are similar to those in other studies seen by the Secretariat. However, they are subject to much uncertainty, and there are grounds for considering that the probability of their being too high is greater than the probability of their being too low. First there is evidence that steel consumption per unit of GDP per capita is falling for the more advanced industrial countries. Second the proportion of steel produced by direct reduction from scrap without the use of coke has been increasing. Third, the amount of coke needed to produce a ton of pig iron has been falling. Because of these factors coking coal demand has tended to fluctuate around a constant trend since the mid 1960s in spite of rapid economic growth up to 1973, and the historical peak demand of 288 million tons (well above the 1976 level of 275 million tons) came in 1970.

Table C-5   **Demand for Metallurgical Coal in the OECD to 2000**
Reference Case
Mtce

	1976	1985	1990	2000
Canada	8	9	10	11
United States	77	88	92	92
OECD Europe	112	117	127	144
Japan	69	93	100	114
Australia/New Zealand	9	12	14	17
OECD Total	275	319	342	378
IEA Total	248	291	312	344

TOTAL OECD COAL DEMAND, SUPPLY AND TRADE

Table C-6 and Annex 1 Table 1-1 provide estimates of total coal demand, supply and trade in the OECD for the reference case. Estimates for the low nuclear sub-case are given for 1985, 1990 and 2000. Estimates for the high nuclear sub-case are given only for the year 2000, since differences from the low nuclear sub-case are small for 1985 and 1990.

Table C-6   **Demand, Indigenous Supply and Trade of Coal in the OECD to 2000**
Mtce

	1976[1]	1985	1990	Low Nuclear 2000	High Nuclear 2000
			Demand		
Canada	24	34	42	57	51
United States	500	769	933	1 052	978
OECD Europe	339	415	485	675	563
Japan	83	115	142	200	170
Australia/New Zealand	40	54	65	98	93
OECD Total	986	1 388	1 667	2 081	1 854
IEA Total	898	1 286	1 547	1 896	
			Indigenous Supply		
Canada	20	40	51	71	65
United States	555	837	1 013	1 181	1 077
OECD Europe	285	321	332	364	364
Japan	20	19	19	19	19
Australia/New Zealand	72	111	155	293	218
OECD Total	953	1 329	1 569	1 928	1 743
IEA Total	859	1 201	1 399	1 624	
			Net Imports (+)/Exports (—)		
Canada	3	—6	—9	—14	—14
United States	—54	—68	—79	—129	—99
OECD Europe	55	94	153	311	199
Japan	60	96	123	180	151
Australia/New Zealand	—31	—57	—90	—195	—125
OECD Total	33	59	98	154	112
IEA Total	42	85	148	273	

1. Demand is not equal to sum of indigenous supply and net imports because of stock changes and statistical discrepancies.

Total OECD coal demand is projected to grow by 3.2 per cent per year from 1976 to the year 2000 in the low nuclear sub-case and 2.7 per cent per year in the high nuclear sub-case. Indigenous supply grows by 3.0 per cent annually in the low nuclear sub-case and 2.6 per cent per year in the high nuclear sub-case.

Coal imports from outside the OECD grow from 33 Mtce in 1976 to 154 Mtce in the year 2000 in the lower nuclear sub-case and 112 Mtce in the higher nuclear sub-case. Japan and OECD Europe both became very large importers. Australia and the United States are projected to become large coal exporters. The pattern of internationally traded coal will go through deep changes in the OECD area. By 1978 the beginning of those changes were already perceptible in the OECD hard coal imports. Annex 2 summarizes the most important developments in the international thermal coal trade which occurred since the 1973/74 oil price rise.

AVAILABILITY OF COAL FROM OUTSIDE THE OECD

Table C-7 displays estimates of net exports and imports by different regions of the world corresponding to the reference case for the OECD. The centrally planned economies are projected to increase their exports, while South Africa rapidly becomes a major exporter of steam coal.

Annex 1 Table 1-3 provides a breakdown of world coal trade between metallurgical and thermal coal. As Figure C-2 below shows, the proportion of thermal coal in international trade (as measured by the sum of net imports of the regions shown in Annex 1 Table 1-3 which are net importers) rises from 25 per cent in 1976 to 61 per cent in the year 2000 in the low nuclear sub-case.

Table C-7   **Projected World Coal Trade**
Reference Case, Low Nuclear

Mtce

	1976	1985	1990	2000
Canada	3	— 6	— 9	— 14
United States	—54	—68	—79	—129
OECD Europe	55	94	153	311
Japan	60	96	123	181
Australia/New Zealand	—31	—57	—90	—195
OECD Total	33	59	98	154
Centrally Planned Economies	—38	—43	—49	— 66
Developing Countries	7	11	2	— 10
South Africa	— 6	—34	—60	— 90
Other	3	7	9	12

1. Net Imports (+) / Exports (—).

THE ENLARGED COAL CASE

As emphasized above, the reference case, based on the continuation of existing energy policies, results in a potential excess demand for oil in 1990 and 2000. This section provides some aggregative estimates of the extent to which OECD oil demands might be reduced by the adoption of additional policy measures to encourage even greater substitution of coal for oil in electricity generation and in other sectors.

The estimates of extra coal demands and supplies in the enlarged coal case, found in Tables C-8 and C-9, are based on judgments taking into account various factors described below. The data are shown in aggregated form because of the uncertainties involved in making projections of this nature and the gradual blurring of the distinctions between steam coal and metallurgical coal and between transformation sector and final consumption demand.

The sorts of policies needed to achieve the enlarged coal case are presented in Chapter B. Some of the policy changes, such as progressive bans on the construction of new oil or gas-fired electricity generating plants, may be difficult to achieve in some cases. The justification for these policies is to prevent upward pressures on energy prices that could result if energy demand and supply trends follow the course indicated by the reference case. The costs imposed by large increases in energy prices are believed to be high relative to the costs of coal substitution.

Electricity Generation

Table C-8 and Annex 1 table 1-4 provide estimates of additional amounts of coal which might be consumed to generate electricity in the main OECD regions in 1985, 1990 and 2000.

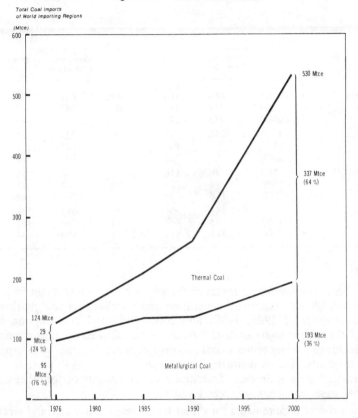

Figure C-2 **World Coal Trade**

In developing the estimates consideration was given to the age structure of oil and gas-fired capacity, and to the existence of dual-fired capacity. Up to 1990, much of the shift toward coal and away from oil and gas in electricity generation is assumed to come from the derating of oil and gas plants, but not necessarily their complete retirement; greater use of coal in facilities with dual-firing capability; the upgrading of existing coal-fired plant and accelerated completion of plants currently planned or under construction; and a few conversions to coal of existing oil and gas plants, primarily in the U.S. where cheap coal makes conversion economics more favourable. Beyond 1990, the increased use of coal in the enlarged coal case is the result of new additions to coal-fired generating capacity, further derating and accelerated retirement of oil and gas plants; and greater use of coal for intermediate and peak load requirements.

The policy changes implied in the enlarged coal case vary from country to country and they partly reflect the time required to improve coal economics as related to infrastructure requirements and the age and composition of the generation mix.

The principal policies include: outright bans on the construction of new oil or gas-fired units over a certain size; subsidies and incentives for accelerated derating, retirement of conversion of oil and gas power plants (including user taxes on oil and gas); a progressive ban on new oil and gas-fired intermediate and peak load units; and progressive limits on oil and gas consumption in dual-fired plants.

They also include a number of attendant policies to resolve environmental constraints, build the required infrastructure, allow greater use of imported coal, and accommodate the required changes in refinery output of heavy fuel oil.

Table C-8   **Thermal Coal Demand in the Enlarged Coal Case**

Mtce

		Low Nuclear			High Nuclear		
		Electricity Generation	Other	Total	Electricity Generation	Other	Total
North America	1985	666	103	769	646	103	749
	1990	849	168	1 017	781	168	949
	2000	913	345	1 258	803	345	1 148
OECD Europe	1985	241	87	328	235	87	322
	1990	294	108	402	281	108	389
	2000	503	221	724	349	221	570
OECD Pacific	1985	56	16	72	52	16	68
	1990	91	25	116	84	25	109
	2000	178	48	226	148	48	196
OECD Total	1985	963	206	1 169	933	206	1 139
	1990	1 234	301	1 535	1 146	301	1 447
	2000	1 594	614	2 208	1 300	614	1 914

Other Uses

Table C-8 also provides estimates of the additional coal that might be consumed by other sectors (outside electricity generation and metallurgical coal markets) in the main OECD regions in 1985, 1990, and 2000. Industrial consumption currently accounts for the largest share of coal for other uses followed by consumption in the residential sector (including miner's coal and district heat), town gas works, patent fuel and briquette plants, and own use by coal mines.

In constructing the estimates of additional coal consumption for other uses in the enlarged coal case, consideration was given to:

— the process requirements for direct heat, steam, hot water and electricity in different industries;

— the adequacy of coal-burning capacity and transportation links to the point of consumption;

— the time required for successful commercialization of new technologies;

— the availability of natural gas and other energy supplies;

— environmental restrictions; and

— past experience with coal consumption and trends in fuel substitution, and the probable demand for new or replacement boilers and furnaces in the future

In the near-term to 1985 and 1990, most of the modest additional rise in coal consumption for other uses shown in the enlarged coal case occurs in the industrial sector, primarily in larger scale under-boiler uses and direct heat applications able to use coal with existing technologies. Furthermore, until 1990 most of the additional consumption is assumed to occur in areas with relatively favourable access to high quality clean coal supplies, and in regions where less-favourable prospects for natural gas create more favourable coal-conversion economics. In other regions and sectors the near-term potential for coal substitution is limited by more favourable outlooks for natural gas supplies, environmental constraints, inadequate coal-infrastructure, technical requirements, and unfavourable coal economics.

Beyond 1990, however, coal consumption for other uses in the enlarged coal case is expected to increase rapidly in most areas as a result of:

— less favourable outlooks for oil and gas supplies and accelerated replacement of oil and gas boilers;

- the successful commercialization of new technologies (coal cleaning and blending; fluidized-bed combustion);
- infrastructure development and expansion of commercial distribution systems; and
- additional opportunities for increased coal consumption in new towns, large-scale residential and commercial building complexes, and industrial parks through numerous applications of district heating schemes and the combined production of heat and electric power.

Furthermore, with the technological advances already indicated, it can be anticipated that coal use in enhanced petroleum recovery will expand, new technologies for producing synthetic gas and liquid fuels will be developed and commercialized, and chemical feedstocks will be produced in greater volume from coal, primarily in the largest coal-producing regions toward the end of the century.

The policy changes implied for successful realization of the additional coal consumption for other uses in the enlarged coal case vary from country to country and over time. They include policies to encourage the substitution of coal for oil both directly, and indirectly through coal substitution for natural gas in certain sectors followed by greater substitution of the freed-up gas supplies for oil in other sectors where the possibilities for direct coal substitution are more limited. The main policy elements, besides the decontrol of oil and natural gas prices assumed in the reference case (but not yet realized in North America), include progressive prohibition of oil and gas use in new or replacement energy-burning facilities capable of using coal; investment incentives for accelerated conversion to coal; progressive user taxes on oil and gas consumption in non-premium direct heat processes and under boiler markets capable of conversion to coal; and accelerated development and commercialization of cleaner, more convenient coal-burning and conversion processes.

These policies will be facilitated by improved co-ordination between energy and regional and urban development planners to promote and facilitate the use of coal-based energy systems in new towns and industrial parks; and by review and resolution of environmental regulations constraining coal-conversion, prompt clarification of existing regulations and, in some cases, preferential allocation of high-quality clean coal supplies. Also important are a number of attendant policies to develop coal distribution networks, allow greater use of imported coal, and promote substitution of freed-up natural gas for oil in markets with limited potential for direct coal substitution.

ENLARGED COAL SUPPLIES

Table C-9 shows the import requirements and the possible sources of the additional coal supplies required to meet the increased demands in the enlarged coal case. The additional supplies originating within the OECD in the enlarged coal case are discussed below.

On the supply side, the production increases over reference case levels required in the enlarged coal case rise from 8 per cent in 1985 to 14 per cent and 26 per cent in 1990 and 2000 respectively. Production increases are shown to occur in all regions. In the year 2000, however, very large increases in production over reference case levels are needed in North America (352 Mtce, or 28 per cent), and OECD Pacific (108 Mtce, or 35 per cent).

European coal imports show significant increases, both absolutely and in percentage terms, over reference case levels equal to 22 Mtce (23 per cent) in 1985, rising to 86 Mtce (17 per cent) in 1990, and to 157 Mtce (50 per cent) in the year 2000. The increases shown would necessitate rapid development of transportation infrastructures on a large scale, the removal of any trade restrictions on coal imports in several European coal-producing nations, and simultaneous increases in domestic coal production greater than those in the reference case. The absolute and percentage increase in Japanese coal imports is significant, but less pronounced than the European increase, and it is concentrated in the period beyond 1990.

These increases in supply and trade cannot be achieved without substantial advance planning and supportive government policies. In general, all of the policies listed in Section B are relevant, but those policies most directly relevant to enlarged coal supplies include: the avoidance of measures which discourage coal imports; government planning of deep harbour facilities for large coal-carrying vessels, and coal-handling transhipment terminals; and the avoidance of any measures to restrict coal exports by coal exporting nationsl. Additional measures to encourage expansion of coal exports production, consumption and trade could also involve: expeditious handling of environmental problems relating to coal production and transportation and avoidance of costly delays in granting leases for coal production; a lowering of severance taxes, possibly offset by higher corporate income taxes or federal loans for community public services; depletion and accelerated depreciation tax provisions for coal comparable to those allowed for oil; and policies to foster competition among coal transport modes (railroads, barges, and slurry pipelines) and among coal producers to ensure that coal prices do not rise unreasonably above costs.

The enlarged coal case would be easier to achieve if combined with high nuclear growth. The results for high nuclear sub-case show a total OECD coal demand in the year 2000 of 2 291 Mtce compared with 2 585 Mtce in the low nuclear sub-case.

Table C-9   **Coal Supplies in the Reference and Enlarged Coal Cases, Low Nuclear Sub-Cases**
Mtce

		Reference		Enlarged	
		Indigenous	Imported	Indigenous	Imported
North America	1985	877	— 74	940	— 74
	1990	1 063	— 88	1 212	— 93
	2000	1 252	—143	1 604	—243
OECD Europe	1985	321	94	335	116
	1990	332	153	350	179
	2000	364	311	400	468
OECD Pacific	1985	130	24	158	19
	1990	174	16	223	6
	2000	312	— 33	420	— 63
OECD Total	1985	1 328	44	1 433	61
	1990	1 569	81	1 785	92
	2000	1 928	135	2 424	162

Reduction in Oil Imports

Achievement of the full potential shown in the enlarged coal case would increase OECD demands for coal over the reference case (low nuclear sub-case) levels by 100 Mtce in 1985, 219 Mtce in 1990 and 504 Mtce in 2000. The increased coal demands would permit corresponding reductions in demands for oil and natural gas (especially in electricity generation). The natural gas thus released could probably be used to substitute for oil. If so the reduction in oil demand, and in oil imports into the OECD, could be 1.4 Mb/d in 1985, 3.0 Mb/d in 1990 and 7.0 Mb/d in the year 2000.

These results imply that the achievement of the enlarged coal case in 1985, without any additional policy measures concerning energy conservation and expanded supplies of other forms of energy over and above those included in the reference case, would allow the IEA's oil import target of 26 Mb/d to be attained. For 1990 the enlarged coal case would reduce the potential gap between world energy demand and supply of 430 Mtoe in the low nuclear sub-case by 35 per cent. For the high nuclear sub-case the gap of 376 Mtoe would be reduced by 39 per cent. For 2000 the gap of 1 260 Mtoe in the low nuclear sub-case would be reduced by 28 per cent, and the gap of 1 185 Mtoe in the high nuclear sub-case would be reduced by 30 per cent.

In conclusion, a strong revival of the OECD's thermal coal industry can perform an important role in helping meet energy demands to the year 2000. But enlarged coal usage alone, it must be said again, is not enough to eliminate potential excess demand. Stronger conservation efforts and more strenuous development of energy supply are also essential.

D

COAL RESOURCES AND RESERVES

Coal resources and reserves are extensive and widely endowed over the world when compared to oil and gas resources, and this dispersion of coal resources enhances the security of coal supply.

Because of the low volume of steam coal moving in international trade, little reliable information concerning global coal resources and reserves was available until the World Energy Conference (WEC) published estimates in 1977. This section is based to a large extent on these WEC estimates.

Coal resources are of course of various average qualities reflecting the complex chemical structure of coal. Key elements of quality are: the calorific value, and the content of sulphur, ash, and bitumen. The different ranges of quality which are greater than those for oil, make estimation and classification on a worldwide scale quite difficult. In this section, definitions by which coal resources and reserves are determined and classified are the same as those used in the WEC report[1].

Estimates of geological resources and technically and economically recoverable reserves may be summarized by continent as follows:

Worldwide geological resources are estimated at about 10 000 billion (thousand million) tons of coal equivalent, while technically and economically recoverable reserves are limited to about 640 billion tons of coal equivalent. This means only about 6 per cent of the total coal resources are envisaged as exploitable under " present" economic conditions (in WEC estimates, mostly conditions prevailing in 1975-76). The recoverable reserves could cover world coal demand at the present consumption level for about 250 years. However, it can be expected that geological resources will be transferred gradually to the recoverable reserves category as energy prices advance faster than coal prices and production techniques improve.

This trend is evident when the current WEC data is compared to the previous data published by the WEC.

Evidently, a re-evaluation has been carried out by many coal producing countries with regard to their coal resources and reserves following the steep rise of energy prices which occurred with the 1973-1974 energy crisis.

---

1. Definitions:
   - *Geological Resources* are understood to mean resources that may become of economic value to mankind, at some time in the future.
   - *Technically and Economically Recoverable Reserves* cover reserves that can be regarded as actually recoverable under the technical and economic conditions prevailing today.

   Norms:
   Hard coal includes anthracite and bituminous coal.
   Brown coal comprises sub-bituminous coal and lignite; the dividing line between these two denominations is the calorific content value of 5 700 kcal/kg (moist, ash free basis); higher contents correspond to hard coal.
   Depth limits:
   Hard coal   1 500 m
   Brown coal   600 m
   Minimum seam thickness
   Hard coal   0.6 metres
   Brown coal   2.0 metres

Table D-1   **World Coal Reserves 1977**

Billion (10^9) tce

	Geological Resources				Tech. and Eco. Recoverable Reserves			
	Hard Coal	Brown Coal	Total	(%)	Hard Coal	Brown Coal	Total	(%)
Africa	173	—	173	(1.7)	34	—	34	(5.3)
North America	1 286	1 400	2 686	(26.6)	122	65	187	(29.4)
Latin America	22	9	31	(0.3)	5	6	11	(1.7)
Asia[1]	5 494	887	6 381	(63.0)	219	30	249	(39.1)
Australia	214	49	263	(2.6)	18	9	27	(4.2)
Europe	536	54	590	(5.8)	95	34	129	(20.3)
Total	7 725	2 399	10 124	(100)	493	144	637	(100)

1. Asia includes European USSR.

Table D-2

WEC Report	(A) Geological Resources (10^9 tce)	(B) Tech. and Eco. Recoverable Reserves (10^9 tce)	B/A %	P/R Ratio[1] (year)
1974	8 603	473	4.9	182
1976	9 045	560	6.2	215
1977	10 124	637	6.3	245

1. World coal production is assumed as 2 597 million tce for 1975.

Public planning for enlarged international coal trade would be advanced by more work in refining estimates of the technically and economically recoverable reserves, both on a national basis as well as on a global basis.

A more detailed country breakdown of technically and economically recoverable coal reserves can be seen in Table D-3.

All the named countries in Table D-3 are presently exporting coal and show a good prospect for expanding coal production. Additionally, countries included as "other" might enter the future international coal market. These include Colombia, Venezuela, and Mozambique. Although these countries are not as well endowed with coal resources as those countries named in the table, most possess coal deposits conveniently situated near the coast or major waterway, providing an important cost advantage in entering the international market.

Finally, it should be noted that the WEC estimates are considered as too low by many coal experts. The WEC report (based essentially on unverified submissions of governments) itself recognized that their estimates were low for many countries due to the limited data available. This is particularly true, and therefore very important, in the case of the United States. The depth limits for coal reserves that the US adopted (300) metres were different from the common norms (1 500 metres). Furthermore, the exploitation factor, i.e., the ratio of actually recoverable reserves to in-place reserves, was assumed as 50 per cent, which is applicable only to room and pillar mining. For longwall mining, open cast mining or new technology such as hydraulic mining, a higher exploitation factor can be recognized. If the introduction of further innovations of exploitation technology, such as underground gasification of coal, or

Table D-3 **Technically and Economically Recoverable Reserves, 1976**
Billion tce (10^9)

Country	Hard Coal	%	Brown Coal	%	Total	%
United States	113	(22.9)	64	(44.4)	177	(27.8)
China	99	(20.1)	—	(– –)	99	(15.5)
USSR	83	(16.8)	27	(18.7)	110	(17.3)
United Kingdom[1]	45	(9.1)	—	(– –)	45	(7.1)
India	33	(6.7)	—	(– –)	33	(5.2)
South Africa	27	(5.5)	—	(– –)	27	(4.2)
Germany	24	(4.9)	11	(7.6)	35	(5.5)
Poland	20	(4.1)	1	(0.7)	21	(3.3)
Australia	18	(3.7)	9	(6.3)	27	(4.2)
Canada	9	(1.8)	1	(0.7)	10	(1.6)
Others	22	(4.4)	31	(21.6)	53	(8.3)
Total	493	(100)	144	(100)	637	(100)

1. The reserve estimates reported for the United Kingdom are judged to be economic to extract only at some future date.

microbiological conversion to gaseous or liquid products, in the future, a large portion of the geological resources would be upgraded to recoverable reserves.

Taking all actual or expected technical and economical changes into account, it is reasonable to say that coal has immense potential for replacing a substantial share of oil in future energy requirements.

Additional information on coal resources and reserves has been included in the country chapters of this report.

E

MARITIME TRANSPORTATION*

Transportation costs are an important component of delivered price for internationally traded coal. This chapter analyses problems associated with the maritime transportation of coal which would expand significantly in both the reference and enlarged coal cases.

Maritime transportation of coal merits special attention because of first, the possibility that the lack of suitable ports and carriers may impede a significant expansion of thermal coal seaborne trade, and because maritime transport costs are and will continue to be a very significant part of CIF costs for imported coal.

The present volume of world seaborne trade in coal, although modest in comparison with petroleum, is important when compared with other commodities: seaborne trade in coal ranks third in volume among dry cargoes after iron ore and grain.

Coal in bulk is shipped in a variety of ship sizes and types. There are a few special vessels known as " colliers" presently used in coal transport. Such vessels are used for coastal or inland waterway transport and some of them have self-discharging capacities. However, for sea routes, ships used are generally conventional dry cargo vessels (twin deckers), bulk carriers and combination carriers (" OBO " or ore/bulk/oil vessels). In 1976 some 10 per cent of all coal seaborne trade was carried out in combination carriers, and 76 per cent in bulk carriers.

Some coastal European traffic has been shipped in tugpushed barges. This method is now used to ship Polish coal to northern French ports and it could have increasing importance in the future if the concept of a receiving port or coal centre for long-distance imports developed in steam coal importing areas, as seems to be the case in Japan.

Starting in the mid-sixties there was a significant shift to use of larger vessels for coal trade particularly on the longer routes. For instance, while 75 per cent of all coal tonnage shipped worldwide in 1965 was carried in ships of under 25 000 dwt[1] and no cargo of over 100 000 dwt took place, by 1976 only 25 per cent of total seaborne trade was in carriers of under 25 000 dwt and over 14 per cent was shipped in vessels of over 100 000 dwt. Increasing Japanese import requirements have been a major reason for this structural change. Vessel size is essentially determined by shipping distances, with larger economies of scale being attained on longer routes. The use of combination carriers (OBO) has been particularly important in the East Coast/Japan trade, the size limit being based on limitations imposed by the Panama Canal. The largest vessels used in coal trade have been between Australia and Japan. There is no record of any coal cargo ship larger than 150 000 tons ever being utilized.

Even though larger carriers are used mainly for long routes, there has been a progressive shift to larger vessels on shorter routes for both iron ore and coal seaborne trade. In fact, it seems that the main reason for the limited extent of such shifts is limitation of port facilities.

---

1.   Dead weight tons.

*   This chapter was written with the assistance of the staff of the Maritime Transport Division of the OECD Secretariat.

FUTURE TRENDS IN COAL SEABORNE TRADE

Ship Sizes

The present trend toward larger ships will accelerate if a significant increase in steam coal trade develops, particularly since a large part of the potential trade will be on longer routes (Australia/Japan, South Africa/Europe) where shifts to larger ships are already occurring.

The limitations imposed by shiphandling facilities at US East Coast ports will limit the size of coal carriers to about 85 000 dwt. Most of the U.S./Japan trade is carried in ships of the "Panamax" size, which refers to ships conforming to the physical limitations of the Panama Canal. Their maximum capacity is around 60 000 dwt although it obviously varies with the type and specific design of the ship. These limitations will persist in the future for that particular route.

A gradual increase in the average size of the vessels used in the coal trade can be expected in the future. No serious technological impediments exist for such a shift, for it has already occurred in the iron ore trade. Furthermore, mid-term prospects of shipyards facing reduced growth rates in both oil and iron ore seaborne trade make it much easier to employ shipbuilding capacity for construction of new coal carriers.

A major problem is estimating the largest bulk carrier size anticipated for coal trade to the year 2000. An upper size limitation of about 200 000 dwt will be imposed by terminal facilities (which will be dealt with below). Given the needs of even a reasonably small imported coal consumer, coal shipments (including transhipments from a main receiving terminal) will be transported in specialized coal carriers of no less than 25 000 dwt. Thus, the cost analysis below is restricted to the 25 000–200 000 dwt range.

Vessel Types

An important fraction of the total trade (about 10 per cent in 1976) is nowadays carried in combination carriers, all of which were OBO's. In the future some coal will still be transported in OBO vessels, particularly the U.S./Japan route[2], but most of the coal will be transported in specialized bulk carriers. The relative importance of the general cargo ship in coal transport will diminish if steam coal is shipped in larger volumes, the reason being that most coal carriers over 25 000 dwt are bulk carriers. It may be presumed, then, that maritime coal transport costs will be determined by bulk carrier traffic. Any advantage of using combination carriers will be translated into local savings on a few routes which should be considered insignificant in the overall trade. Costs, therefore, are estimated for bulk carriers only.

Only diesel motor ships using fuel-oil are considered in the cost analysis because of their favourable economics compared to steam turbines using the same fuel. This assumption could be subject to criticism, however. It could be that as the lighter cuts of the barrel become more valuable, the heavier end of the barrel will become more residual, and vessels with steam turbines could use that residual to reduce costs. However, diesel motor ships are assumed to have the lowest transport costs in the late 80s and early 90s.

Coal Ports and Terminals

Port facilities will impose an upper limit on the future vessel sizes used in coal seaborne trade. The limitations are both in the form of berths and loading/unloading

---

2. Traditionally combination carriers have taken crude oil from the Persian Gulf to the US East Coast, coal to Japan via the Panama Canal and transit in ballast from Japan to the Gulf.

facilities. The iron ore trade has led in development of new, modern port facilities which can also be utilized in coal trade. Since 1971, facilities have existed which could handle up to 10 000 tons per hour of iron ore. Modern coal loading ports (Roberts Bank and Richards Bay) can attain loading rates of 35 000 tons a day for 120 000 dwt vessels and above. However, even at advanced coal terminals, typical discharging rates are lower (25 000 tons per day for the most modern coal berths at Rotterdam, and Ensted for example).

Bulk commodities like coal are more costly to handle than fluids like oil because the first requires berths and cranes while the second may be pumped aboard at a berth or via a pipeline and buoy. Coal traders are generally not optimistic of cost reduction in bulk handling in the future. For the purpose of the cost estimates in this chapter, however, some allowance is made for improved technology in coal handling by assuming that even in loading and discharging coal in larger coal ships, the present average port time of 6 days per round-trip passage will be held constant.

MARITIME TRANSPORT COST

In general, maritime transportation services are the responsibility of the coal buyer, the contract terms being on a FOB or FAS (free alongside ship) basis.

Various shipping arrangements are generally available for the transport of coal. One type is a spot charter, usually called voyage chartering, whereby shippers contract the service of a vessel to transport a given quantity of coal between two ports. Generally, however, seaborne coal trade is carried out under time-chartering contract, the time ranging from several months to the whole life of a ship. The type of shipping contract chosen usually parallels coal supply conditions. Spot or short-time shipping rates are variable; a temporary excess demand tonnage may cause freight rates to rise sharply, a slump in demand may cause them to fall as sharply — even to a point of not covering capital costs of the ship owner.

Table E-1 **Maritime Transport Cost**
Main Assumptions for Bulk Carriers

	Ship Size (dead weight tons)					
	25 000	50 000	75 000	100 000	150 000	200 000
Annual Operating Costs (1976 $ million)	365	270	225	205	190	185
Annual Operating Costs (1976 $ million)						
— in 1985	1.2	1.45	1.6	1.75	2.0	2.8
— in 2000	1.4	1.65	1.8	1.95	2.2	3.0
Service Speed (knots)	14.5	15.25	15.75	16.1	16.4	16.5
Daily Consumption (tons)						
— at sea	30	52	72	85	105	125
— at port[1]	3	3	3	3	3.5	3.5
Days in port per round voyage	6	6	6	6	6	6
Bunker price (1976 $/ton)						
— 1985	90	90	90	90	90	90
— 2000	130	130	130	130	130	130

1. Ships do not have self-discharging capabilities.

Long-term coal supply contracts generally include long-term shipping arrangements. Even though there is little published information on the nature of such contracts, they are known to provide the owner with a return on the capital plus recovery of operating costs usually incorporating escalation clauses. Such contracts cover 95 per cent of the present total seaborne coal trade under long-term coal sale agreements to steel companies. This practice will dominate the enlarged seaborne coal trade foreseen in this report and long-term average ocean freight rates will determine the seaborne costs of total delivered cost of coal in importing areas.

COST CALCULATIONS

Seaborne

To recapitulate, in estimating long-run transport costs per ton of coal, some operating and capital cost assumptions have to be made. These assumptions are summarized in Table E-1.

The seaward costs of maritime transportation are composed of:

a) *Capital costs of the vessels.* For the larger ships, capital costs per dwt are extrapolations of present conditions since no 200 000 dwt bulk carriers have been built as yet. They are assumed to remain constant in real terms to 2000.

b) *Running costs* (excluding fuel). These costs comprise crew's wages, costs of repairs and maintenance and insurance. Salaries are assumed to increase in real terms by 2 per cent p.a. from 1985 to 2000.

c) *Fuel costs.* These costs are a function of fuel consumption and fuel price. Fuel price is assumed constant through 1985 and rising at 2.5 per cent p.a. in real terms through 2000.

In addition to the assumptions expressed in Table E-1 a number of operating conditions will be presumed as follows:

1. *Ship life:* 18 years in all cases.
2. *Capital charges:* estimated at 10 per cent DCF, same conditions as those used in other sections of the report to estimate cost of capital to public utilities.
3. *Engine type:* All vessels are diesel motor ships using fuel oil.
4. *Down-time:* Ships are out of work for repair or dry-docking operation an average of 20 days/year.
5. *Round voyage* (laden and ballast): Five representative routes were chosen:
   a) 26 000 nautical miles (NM): Australia-Western Europe;
   b) 20 000 NM: East Coast North America-Japan or West Coast North America-Western Europe via the Panama Canal (no size limitations or canal tolls have been included in Table E-2;
   c) 14 000 NM: South Africa-Japan or South Africa-Western Europe;
   d) 7 000 NM: US East Coast-Western Europe;
   e) 2 000 NM: China-Japan or Inter-European trade.

The corresponding costs in 1976 dollars per ton are shown in Table E-2. They include only seaborne costs, and represent cost conditions of 1985 and 2000. The economies of scale in using larger ships are evident for the longer routes. For instance, in the Australia-Europe route transport costs using a 100 000 dwt vessel are half those of a 25 000 tonner.

Table E-2 **Seaborne Transport Costs in 1985 and 2000**
In 1976 U.S. $/ton

Ship size (dwt)	Year	Length of Round Trip (Nautical Miles)				
		26 000	20 000	14 000	7 000	2 000
25 000	1985	29.7	23.2	16.8	9.2	3.8
	2000	35.2	27.5	19.8	10.8	4.3
50 000	1985	20.5	16.0	11.5	6.3	2.6
	2000	24.5	19.0	13.6	7.4	2.9
75 000	1985	16.5	12.9	9.3	5.1	2.1
	2000	19.7	15.4	11.1	6.0	2.4
100 000	1985	14.2	11.1	8.0	4.4	1.8
	2000	16.9	13.2	9.5	5.1	2.0
150 000	1985	11.8	9.2	6.6	3.6	1.5
	2000	13.9	10.9	7.8	4.2	1.7
200 000	1985	11.3	8.8	6.4	3.5	1.5
	2000	13.1	10.3	7.4	4.0	1.6

Additional Costs

In addition to maritime transportation costs, transport cost estimates should also include shore costs such as loading and unloading costs, including port fees. Furthermore, the expected increase in world seaborne steam coal trade will require expansion of port capacities at both importing and exporting areas to increase throughput capacity beyond present levels and to allow use by larger vessels, particularly in the longer routes where transport economies of scale for larger vessels are evident as shown in Table E-2 above. Capital costs for new port construction or existing port expansion vary widely for different locations, the main cost items being the amount of dredging needed and new berth construction. In Japan the cost of construction of a new coal centre has been estimated at U.S.$125 million for a yearly throughput capacity of 7.5 million tons and U.S.$250 million for 15 million tons per year. The coal centre will be able to service vessels of up to 100 000 dwt. In Australia, the cost of expanding the port of Newcastle to receive 120 000 tonners has been estimated at some $87 million.

Port Dues

In general, ports are developed and managed by local, provincial or national authorities, especially as port fees do not necessarily correspond to good cost recovery practice. A recent OECD survey of port fees found them as low as $0.05/dwt in a typical U.S. harbour and as high as $0.65/dwt in a typical European port (both values refer to 1976). Thus, it is difficult to estimate any future trend in port fees. Nevertheless it is assumed that in the future, docking fees will amount, on an average, to $0.7/ton of coal for each round voyage.

Loading/Unloading Costs

Bulk terminals are usually run and costed on a commercial basis. These facilities are fashioned to the requirements of the site. For example, where high tides and strong winds are common, capital costs may be much higher than in calmer conditions.

A recent analysis submitted to the IEA estimated representative costs as $0.28 per ton for loading and $0.55 per ton for unloading. However, a survey carried out by the Secretariat found that costs are between $1.00 and $1.50 per ton, depending on the port and the type of unloading arrangement (transhipment to barge or coaster, or

Figure E-1 **Economies of Scale in Maritime Transport**
(Including terminal charges)

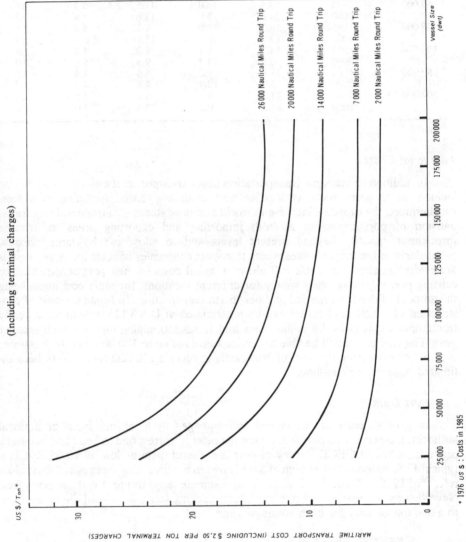

* 1976 *US* $: Costs in 1985

direct transfer to stocking area). Loading rates in Europe tend to be higher than $0.28 per metric ton. Loading charges and port fees are particularly high in Australian ports where they are as much as $4.50 per ton, but it is believed on the average they are considerably lower.

From these various sources it can be estimated that loading and unloading costs through 2000 will amount to $1.80 per ton in average.

To summarize, a total of $2.50/ton should be added to the seaward costs calculated in Table E-2 to give an estimate of the total overall maritime transportation costs. The resulting figures should be considered as approximate orders of magnitude for the maritime component of internationally traded steam coal. Figure E-1 indicates the economies of scale in seaward transportation (including terminal charges) for different round distances and ship sizes.

Transhipment

The very different transportation conditions in prospective steam coal importing countries make it very difficult to estimate the cost of transportation from the receiving end of the seaborne transport leg to the consuming point.

For power stations located on the coast with port facilities, the use of small size vessels (thus avoiding expensive port expansion) from a coal transhipment port could be an attractive option. This seems to be the case in Japan where a coal centre is projected to use the economies of scale in long-distance seaborne transportation. For the future shipment of steam coal into Europe, several ports can be identified which could serve as coal receiving and distribution centres, especially in Northern Europe and the Mediterranean area. International arrangements could be made to allow these centres to serve coal importing needs in different countries. For no more than $5 per ton coal could be transhipped from the closest of these coal centres to almost any coastal location in Europe or Japan. Transhipment could be done with small size bulk carriers or seaward, large capacity barges.

Using land transport rather than seaborne transhipment, coal centres could serve a smaller area at the same additional transport cost. A cost of $5 per ton should, on the average, allow the coal to reach 200 km using railways of 400 km using an existing inland waterway network.

F

ECONOMICS OF FUEL SUBSTITUTION BY COAL

Coal's share of the total primary energy requirements in the OECD fell from 26 per cent in 1960 to 20 per cent in 1976. Furthermore, the actual volume of coal consumption rose by only 46 Mtce during this period as the increases in coal consumption by the electricity generation sector and the iron and steel industry were nearly offset by declines in the remaining sectors. The prospects for reversing this decline will depend on the relative prices, efficiencies, and available supplies of different fuels, government energy policies, and the successful resolution of several technical and environmental constraints that presently limit the possibilities for coal substitution.

Not all fuels are perfect substitutes for one another and there are considerable short-term rigidities in the established patterns of energy consumption. Coal is difficult and expensive to distribute, and the need to control dust, dispose of ashes, allocate storage space, and burn coal in an environmentally acceptable manner makes coal inconvenient and costly for all but the largest users. Furthermore, conversion of the existing stock of capital equipment from oil and gas to coal will be more difficult and time consuming, for technical, economic and environmental reasons, than the conversion away from coal that took place in the past.

Nevertheless, the prospects of coal have been greatly enhanced by recent fundamental changes in the world's energy markets. Higher oil prices since 1973, the increasing cost and delay in nuclear power programmes, and the prospect of limited oil and (at least locally) natural gas supplies pushing prices further upward in the future, have increased coal's economic attractiveness. The following sections explore the probable future relationship between coal and oil prices, the competitive position of coal vis-à-vis alternative fuel supplies and the near-term and long-term possibilities for coal substitution in each of the principal energy-consuming sectors.

INTERFUEL COMPETITION IN ELECTRIC POWER GENERATION

The electricity generation sector is composed of a relatively small number of very large users of primary energy who have an almost continuous need for new or replacement capacity. This is especially true since the growth of electricity demand has been and is expected to remain high relative to that of overall energy demand in the future. In addition, since electricity can be generated in a wide variety of centralized or decentralized power stations using coal, oil, gas, and nuclear fuels, hydro power and geothermal resources, the electricity sector could be expected to respond more readily to the substitution possibilities that arise as a result of changes in the relative prices and supplies of different fuels. Furthermore, since electricity competes with other fuels in certain end-use applications (space and water heating, cooking, mechanical drive, etc.), the substitution possibilities afforded by the electricity generation sector at the primary input level are enhanced by the substitution possibilities afforded by electricity itself. In the near-term, therefore, the electricity generation sector will continue to be an extremely important medium for fuel substitution. Coal, because of its inconvenience and other undesirable qualities, could play a much smaller role in fuel substitution without electricity.

COMPARATIVE ELECTRICITY GENERATION COSTS

The electricity generation sector has been the object of research for many years and a good deal of technical information describing the various means of electric power generation is readily available. Nevertheless, the factors affecting the relative economics of alternative electricity generation systems remain subject to a large number of uncertainties about which there can be honest differences of opinion. New power plants in the early stages of planning or construction, for example, will not enter commercial operation until the mid-1980s, and their useful lifetime may extend for a decade or more into the next century. With energy markets still in disequilibrium, estimates of electricity generation costs in new power plants over such a long time horizon are based, by necessity, on a number of technical, economic, and institutional assumptions about the uncertain course of future events:

—   The capital costs of new power stations have escalated very rapidly in recent years and they have been significantly underestimated in the past;
—   The competitive merits or commercial feasibility of new technologies, often based on cost estimates constructed from " paper " plants or small scale pilot projects, remain unknown;
—   Future changes in environmental and safety regulations, and in social, economic and energy policy objectives are difficult to assess, but they could easily shift a country's preference for particular fuels;
—   The choice of fuels used for electricity generation can have simultaneous repercussions on the prices of competitive generating fuels and on the competitive position of electricity and other energy supplies in certain end-use markets: a widespread shift to coal for electricity production, for example, could affect the prices of the nuclear, oil and gas fuels it replaces.

Many estimates of the comparative costs of electricity generation in new coal, oil and nuclear power stations have been published in the last few years, and a large number of these estimates were collected and analysed in order to assess the competitive position of coal-based electric power. The estimates reviewed come from a variety of public and private sources, and care was taken to include estimates covering the major OECD regions and countries. Most of the estimates reviewed were constructed during the last two years, and they pertain only to the cost of building and operating *new* power stations intended for initial start-up during the mid-1980s.

In view of the uncertainties mentioned above, there was considerable variation in the basic assumptions underlying the estimates reviewed. In addition, some of the estimates were project-specific, referring to individual plants and specific sites, while other estimates were generalized. Comparability was limited even further by the use of different accounting procedures.

In order to develop comparative measures of electricity generation costs in new nuclear, coal and oil-fired power stations, the available estimates were adjusted to conform to a common set of definitions, assumptions, and accounting practices. The representative estimates of electricity generation costs in alternative power plants presented in this report were derived from this exercise[1].

The cost estimates used in the following analysis are presented solely as a benchmark to illustrate current estimates of the likely cost of new coal, oil and nuclear power stations under construction or being planned for commissioning in the mid-1980s. As such, the cost estimates abstract from all items which depend exclusively on the general regimes in force in each country. For example, no attempt has been made

---

1.   A technical annex describing the accounting procedures used in the construction of the cost estimates presented here is available upon request from the Secretariat.

to account for the specific policies of individual regulatory commissions governing the financial and fiscal considerations that determine a utility's rate base, capital charges, and allowable rate of return. Neither do these generalized cost estimates include assessment of a variety of special conditions, determined by local circumstances, that can easily influence the design and cost of individual power stations, such as land costs, taxes etc.

The cost estimates presented below refer only to the cost of electricity at the point of generation. They should not be confused with the overall cost of supplying electricity to final consumers which includes transmission and distribution costs, certain overhead expenses and the particular design of the electricity tariffs for different classes of consumers.

Subject to the limitations mentioned above, Table F-1 presents comparative electricity generation cost estimates in new coal, oil and nuclear power stations. All costs are presented in 1976 U.S. dollars in constant money terms, and were normalized to a 1986 operating date in order to account for differences in the construction periods for each plant and the interest charges on funds used during construction. The capital cost estimates include all direct and indirect expenses incurred during construction plus a contingency to account for unforeseen expenses, possible construction delays, and initial start-up costs.

The capital cost estimates do not include any allowance for escalation in the cost estimates due to changes in real prices[2]. Neither do they include land costs or other cost elements not directly related to the power station that depend on general policies in each country such as dues, taxes, insurance payments, and licensing criteria. These items have been excluded from the general analysis presented here because of the wide divergencies from country to country (and site to site) in inflation rates, wages, the prices of land, construction materials, and power plant equipment, and other important parameters such as exchange rate changes.

Regardless of the actual lifetime of the power stations, the capital charges per kwh at the different operating rates shown in Table F-1 are based on a capital recovery period of 20 years and a constant money (real) discount rate of 10 per cent. In international comparisons, however, it is impossible to choose a single "best" value for a constant money discount rate that will adequately reflect the considerable disparities in the capital markets and sources of utility financing that exist between countries. Constant money discount rates in use today vary from less than 5 per cent to more than 10 per cent in OECD countries. For this reason, the results of a sensitivity analysis showing the impact on capital costs per kwh for various power plants for each 1 per cent reduction in the discount rate is also presented in Table F-1. The operation and maintenance costs per kwh are typical for new plants commissioning in the mid-1980s, and they include no allowance for real escalation of these costs beyond 1986.

The nuclear capital cost estimate shown in the table ($700 per kw) corresponds to early 1976 licensing requirements in the United States, ideal site conditions, and construction by an experienced contractor. Different licensing requirements, reactor systems, environmental and safety regulations and the site-specific factors mentioned above could cause nuclear power plant investment costs to vary considerably from one location to another within a probable range of $600 to $1 000 per kilowatt (kw) of installed capacity in 1976 dollars.

---

2. In other words, the capital costs of constructing new power stations are assumed to escalate at the same rate as the general price level in *current money terms*. However, since all costs in this analysis are presented in constant money terms, no allowance for general inflation needs to be included.

The nuclear fuel-cycle cost estimates are based on a uranium (U_3O_8) price assumption of $35 per lb ($77 per kilogram) in 1976 dollars for uranium delivered in 1986, and no allowance for real escalation beyond 1986. The nuclear fuel-cycle cost estimates are believed to be typical for a new pressurized light water reactor without reprocessing of the spent fuel elements[3].

The two estimates of electricity generation costs in oil-fired power stations presented in Table F-1 refer to (*i*) a conventional oil-fired power station without a flue gas desulphurization (FGD) or scrubber unit burning low-sulphur heavy fuel oil only; and (*ii*) an oil-fired power station with a flue gas desulphurization unit (with 100% FGD) burning high-sulphur heavy fuel oil only. The price assumptions for high and low-sulphur heavy fuel oils shown in the table are both consistent with the basic assumption of this report that crude oil prices rise by 2½ per cent per year in real terms beyond 1985. The current depressed market for heavy fuel oils is expected to disappear as refineries gradually adjust the composition of their output toward the lighter end of the barrel. The differential between high and low-sulphur heavy fuel oils shown in the table (approximately $15 per t.o.e.) is roughly equal to the cost of desulphurization at the refinery which is, at present, slightly lower than the cost of desulphurization at the power plant.

The three estimates of the cost of electricity generation in coal-fired power plants presented in Table F-1 refer to:

i) 　a conventional coal station designed to burn low-sulphur coal without a scrubber unit (without FGD);

ii) 　a coal-fired power station designed to burn high-sulphur coal with a scrubber unit capable of cleaning 100 per cent of the flue gas (with 100 per cent FGD); and

iii) 　a coal plant with a smaller scrubber unit designed to clean 50 per cent of the flue gas (with 50 per cent FGD).

The capital cost estimates refer to coal plants designed to burn hard coal only, and they include the costs of electrostatic precipitators (coal plants burning lignite, brown coal, and certain other coals with low-sulphur and high ash contents may cost an additional $30-50 per kilowatt to cover the cost of more expensive ash collection systems). The flue gas desulphurization costs refer to the lime or limestone wet slurry process and they include the cost of sludge fixation and disposal[4].

The coal price shown in Table F-1 is roughly equal to the current weighted average price of steam coal delivered to utilities in the OECD area (in 1976 dollars), and it includes an allowance for escalation in the real price of steam coal equal to 1 per cent per year beyond 1985. The weighted average coal price shown in the table is included solely as a reference point to illustrate, in a broad manner, the overall competitive position of coal-fired electricity generation in the OECD area today. It is *not* a realistic figure on which individual consuming countries could assess the comparative cost of coal in relation to other fuels in the long term.

The single coal price shown in the table has relatively little significance for some countries (or for individual power stations) since one price can never reflect adequately the wide range of (*i*) regional coal prices; (*ii*) differences in the costs of surface and deep-mined coal; (*iii*) differences in the prices of domestic and imported steam coals;

3.　The nuclear fuel cycle cost estimates are described in greater detail in an annex available from the Secretariat (see footnote 1 on page F-3).

4.　The FGD cost estimates presented here are based on the high-end of the range of base capital costs for FGD systems presented in the environmental section of this report (Section G). The FGD costs presented in Table F-1 also include interest on funds during construction and an energy penalty to offset the electricity consumed by operation of the scrubber unit.

and (*iv*) differences in coal prices related to differences in coal quality. For example, the actual prices of steam coal delivered to utilities in the OECD area included in the weighted average price shown range from a low of $ 10-20 per t.o.e. (10^7 kcal) for some mine-mouth power stations in North America, Australia, and Germany (burning lignite) to a high of $ 70-90 per t.o.e. in several European countries burning the highest-cost domestic coal supplies.

Since the weighted average coal prices shown in the table cannot reflect the wide differences in the prices and qualities of all the coal types likely to be available in specific countries in future years, the following sections begin with an examination of the relationships shown in Table F-1 and then proceed to several alternative analyses of the probable competitive position of coal vis-à-vis oil and nuclear power for electricity generation. It should be borne in mind, however, that these cost comparisons are based largely on the world market price of coal, and a different result would be obtained in markets where coal prices are well above or below world market levels.

COAL'S COMPETITIVENESS WITH OIL

In Table F-1 it is shown that on an averaged OECD-wide basis new coal-fired power stations with expensive FGD facilities could (when they begin operation in 1986) produce electricity at a lower cost than oil-fired power stations with similar environmental protection devices at approximately today's oil price. Furthermore, in the setting of an assumed rise in oil prices by 2 ½ per cent per year beyond 1985, the comparative economic advantage of coal-fired electricity generation will become more pronounced: at an operating rate of 5 500 hours per year (63% capacity factor), the table shows that the coal plant with full desulphurization facilities could produce electricity at an average total cost during the first 20 years of operation almost 35 per cent less than the average cost of electricity produced in an oil-fired plant over the same period[5].

Furthermore, even though the higher capital cost of coal-fired power plants work to coal's competitive disadvantage as operating rates fall, the coal-fired plants still maintain a competitive advantage over the oil plants at the lowest operating rates shown in the table.

Maximum Competitive Coal Prices

Since the weighted average coal price has little significance for specific power plants, an alternative analysis was made to determine the maximum competitive delivered prices utilities could afford to pay for high and low-sulphur steam coals and still generate electricity at the same overall cost per kwh as oil-fired power stations with similar environmental protection devices. The analysis is based on the total cost of electricity generation in the oil-fired power plants, and on the capital and operation and maintenance costs of the coal plants with and without FGD systems presented in Table F-1.

Table F-2 presents the maximum competitive or "breakeven" prices utilities could afford to pay for (*i*) low-sulphur coals burned in power plants without FGD facilities and still generate electricity at the same cost as oil-fired power plants burning low-sulphur heavy fuel oil without FGD; and (*ii*) high-sulphur coals burned in power

---

5. Since oil prices are assumed to rise by 2 ½ per cent per annum after 1985 and coal prices by 1 per cent, the actual cost differential between coal and oil-fired electricity production would rise from 8.8 mills/kwh or 25 per cent in 1986 to 16.7 mills/kwh or 37 per cent in the year 2000.

Table F-1   Cost Estimates for Electricity Generation in New Baseload Nuclear, Oil, and Coal-Fired Power Stations

In 1976 dollars; Average Costs per kwh during the first 20 years of operation for new plants commissioning in 1986

Average Cost per kwh over first 20 years ($ mills)	Nuclear	Fuel Oil[1]		Bituminous Coal[2]		
	PWR 2 x 1 100 MW	2 x 600 MW		2 x 600 MW		
		Low-Sulphur HFO conventional	High-Sulphur HFO with 100 % FGD	Without FGD	With 100 % FGD	With 50 % FGD
Capital Cost	14.9	7.5	9.6	9.6	12.4	11.0
Operation and Maintenance Cost	2.4	2.0	4.2	2.2	5.1	3.6
Fuel Cost	6.5	31.0	29.0	10.8	11.3	11.1
Total Average Cost per kwh at 5 500 h/a	23.8 (22.8)[3]	40.5 (40.0)	42.8 (42.2)	22.6 (22.0)	28.8 (28.0)	25.7 (25.0)
7 000 h/a	20.7	38.9	40.8	20.6	26.1	23.3
6 000 h/a	22.6	39.9	42.0	21.8	27.8	24.8
5 000 h/a	25.3	41.2	43.8	23.6	30.0	26.8
4 000 h/a	29.5	43.3	46.4	26.2	33.4	29.8
3 000 h/a	36.3	46.7	50.8	30.6	39.1	34.9
Cost of Construction ($/KW)	$700	$350	$450	$450	$580	$515
Average Fuel Cost for the period 1986-2006 :						
$ per toe (10^7 kcal)	$25.79	$132	$118	$45	$45	$45
$ per 10^6 Btu	$ 0.65	$3.33	$2.97	$1.13	$1.13	$1.13
Heat Rate :						
Btu/kwh	10 000	9 325	9 715	9 500	9 880	9 690
kcal/kwh	2 620	2 350	2 445	2 395	2 490	2 440
Memorandum Item :						
Total Cost per kwh 1986	23.5	34.8	36.4	21.5	27.6	23.5
at 5 500 h/a in : 1990	23.9	37.4	38.8	21.9	28.0	23.9
2000 ($ mills)	24.9	45.2	45.8	22.9	29.1	24.9

Notes :

Assumptions common to all plants :
i) All costs are expressed in 1976 $ U.S. constant money, 1 mill = 10^{-3} $ U.S.
ii) All plants are assumed to begin operation in 1986.
iii) The capital cost estimates (capital charges) are based on a real discount rate of 10% over a capital recovery period of 20 years (10% at 20 years = 11.746% p.a.).

Notes :
1. Both of the oil prices shown in the table reflect the assumption that crude oil prices rise by 2.5% p.a. after 1985. The difference between the two oil prices ($15 per toe) reflects a quality differential between high and low-sulphur heavy fuel oils, and it is roughly equal to the cost of desulphurization at the refinery.
2. The coal price shown in the table ($ 3 2 per tce) is based on the weighted average price of steam coal delivered to utilities in the OECD area and real coal price escalation of 1% p.a. after 1985, and it is presented here solely for the purpose of illustration. The wide regional and quality-related differences in coal prices will affect coal's competitive position in individual countries See text.
3. Figures in parenthesis are the alternative estimates based on a real discount rate of 9% instead of 10% in the computation of capital charges; thus one percentage point decrease of discount rates leads to approximately 4% decrease of total average cost for nuclear, 1% for oil, and 3% for coal, respectively.

h/a : hours per year
FGD : flue gas desulphurization

Table F-2 **Maximum « Break-even » Prices for High and Low-Sulphur Steam Coal Relative
to Oil in New Power Plants**

1976 $ U.S.

	Operating Rate hours/year	Maximum Competitive Coal Prices[3]	
		Low-Sulphur Coal-without FGD	High-Sulphur Coal with FGD
In Mills/kwh[1]	7 000	29.1	26.0
(1 mill = 10[-3] $ U.S.)	5 000	28.4	25.1
	3 000	26.9	23.0
	7 000	3.06	2.63
In $ per 10[6] BTU[2]	5 000	2.99	2.54
	3 000	2.84	2.33
In $ per metric ton of oil equivalent	7 000	121.79	104.36
($ 10[7] kcal)	5 000	118.64	100.79
	3 000	112.69	92.45
In $ per metric ton of coal equivalent	7 000	85.25	73.05
($/0.7 × 10[7] kcal)	5 000	83.05	70.55
	3 000	78.88	64.72

1. Competitive coal prices per kwh are equal to the total cost of generating electricity with oil *minus* the non-fuel costs (capital + operation and maintenance costs) of the coal plants at the different rates of utilization assumed above.
2. At the heat rate assumptions presented in Table F-1, i.e. 9 500 Btu/kwh and 9 880 Btu/kwh in coal plants without scrubbers and with scrubbers, respectively.
3. Delivered to utilities.

plants with 100 per cent FGD and still generate electricity at the same cost as oil-fired power stations burning high-sulphur heavy fuel oil with 100 per cent FGD.

If utilities were able to buy steam coal at the maximum competitive or " breakeven" prices shown in Table F-2, they would be indifferent in economic terms between the choice of coal and oil for electricity generation. However, if coal were available to utilities at prices less than the maximum breakeven prices shown in the table, there would be an economic advantage working in coal's favour and the cost of electricity generation in coal-fired power plants would be less than the cost of electricity generation in oil-fired plants. This breakeven analysis, therefore, presents the upper bound for coal prices if coal is to be competitive with oil in electricity generation in new power plants.

For example, at an operating rate of 5 000 hours per year, Table F-2 shows that coal will be competitive with oil for electricity generation in power plants subject to similar environmental control regulations as long as the delivered prices for high and low-sulphur coal are equal to or less than $ 101 and 119 per t.o.e., respectively[6]. This analysis also shows that, on a heat equivalent basis, the maximum competitive delivered prices for high and low-sulphur coal must be $ 17 and $ 13 per t.o.e., respectively, below the delivered prices of high and low-sulphur fuel oils to offset the higher capital and operation and maintenance costs of the coal plants at an operating rate of 5 000 hours per year. These differentials, on a heat equivalent basis, are greater at lower operating rates.

The analysis of maximum competitive prices presented above is based on comparisons between:

6. $ 71 and $ 83 per t.c.e., respectively.

    i)   Low-sulphur coal and low-sulphur heavy fuel oil burned in power plants without FGD; and

    ii)   High-sulphur coal and high-sulphur heavy fuel oil burned in power plants with 100 per cent FGD.

Other comparisons, for example between oil-fired power plants burning high-sulphur heavy fuel oil without FGD facilities and coal-fired power plants with 100 per cent FGD capability, would widen the differential required between coal and oil prices on a heat equivalent basis, but they are of less relevance to this analysis since such comparisons imply that coal-fired power stations would be subject to more restrictive sulphur emission controls than oil-fired power stations.

COAL'S COMPETITIVENESS WITH NUCLEAR POWER

The competitive position of coal-fired electricity generation vis-à-vis nuclear power in the OECD is less definitive than the choice between coal and oil. At the capital and fuel costs of coal and nuclear power stations shown in Table F-1, only coal plants capable of burning low-sulphur coal without FGD systems appear to have a distinct advantage over the cost of nuclear power in baseload operations. However, as operating rates fall, the higher capital cost of nuclear power stations work to nuclear power's competitive disadvantage, and the coal plant with 50 per cent FGD becomes competitive with nuclear power at operating rates lower than 4 000 hours per year.

The estimates for coal and nuclear electricity generation costs shown in Table F-1 illustrate an important structural difference between the cost of coal and nuclear power. Nuclear power costs, on the one hand, are dominated by fixed or capital costs while coal-fired electricity costs are dominated by fuel costs. Since the table presents only one capital cost estimate for nuclear power plants and only one coal-price, despite the wide range of current costs and future estimates that exist for both, definitive conclusions about the cost of coal vis-à-vis nuclear power cannot be drawn from the information presented. For example, with the same nuclear fuel and operation and maintenance costs shown in Table F-1, a plausible range of construction costs for nuclear power stations of $ 600 to $ 1 000 per kw of installed capacity would produce a range of total electricity generation costs of 20.7 to 30.3 mills per kwh for new nuclear power stations.

Maximum Competitive Coal Prices

An alternative analysis was performed to determine the maximum competitive or "breakeven" prices for high and low-sulphur steam coals that utilities could afford to pay and still generate electricity at the same cost per kwh as nuclear power plants (based on the total nuclear power costs and the capital and operation and maintenance costs for coal plants shown in Table F-1). Table F-3 shows, for example, that coal-fired electricity generation will be competitive with nuclear power at an operating rate of 5 000 hours per year as long as the delivered prices of high and low-sulphur coals are equal to or less than $ 19 and $ 36 per t.c.e., respectively[7]. At an operating rate of 3 000 hours per year, however, coal will be competitive with nuclear power at delivered prices up to $ 24 and $ 48 per t.c.e. respectively for high and low-sulphur coal[8]. It appears, therefore, that coal will be competitive with nuclear power in baseload operation in the lower cost coal mining areas of the OECD and possibility in coastal areas with easy access to low-sulphur imported steam coals. In other areas,

---

7. $ 27 and $ 52 per t.o.e. respectively.
8. $ 34 and $ 69 per t.o.e. respectively.

Table F-3   **Maximum « Breakeven » Prices of High and Low-Sulphur Steam Coal Relative to Nuclear in New Power Plants**

1976 $ U.S.

	Operating Rate hours/year	Maximum Competitive Coal Prices[3]	
		Low-Sulphur Coal without FGD	High-Sulphur Coal with FGD
In Mills/kwh[1]	7 000	10.9	5.9
1 Mill = 10⁻³ $ U.S.	5 000	12.5	6.6
	3 000	16.5	8.5
In $ per 10^6 BTU[2]	7 000	1.15	0.60
	5 000	1.32	0.67
	3 000	1.74	0.86
In $ per metric ton of oil equivalent	7 000	45.53	23.70
($/$10^7$ kcal)	5 000	52.21	26.51
	3 000	68.92	34.14
In $ per metric ton of coal equivalent	7 000	31.87	16.59
($/$0.7 \times 10^7$ kcal)	5 000	36.55	18.56
	3 000	48.25	23.90

1.   Competitive coal prices per kwh are equal to the total cost of generating electricity in nuclear power stations *minus* the non-fuel costs (capital + operation and maintenance costs) of the coal plants at the different rates of utilization shown above.

2.   At the heat rate assumptions presented in Table F-1, i.e. 9 500 Btu/kwh and 9 880 Btu/kwh in coal plants without scrubbers and with scrubbers, respectively.

3.   Delivered to utilities.

there may be a considerable market for coal in intermediate load operations or in smaller decentralized power stations.

It is important to note, however, that the capital costs of coal and nuclear power stations used in the breakeven analysis do not include any allowance for real escalation of these costs over and above the general price level. Because the capital cost of nuclear power stations is considerably greater than that of coal plants, an equal rate of escalation in the real cost of constructing both coal and nuclear plants would widen the capital cost difference in absolute terms, thereby allowing the maximum breakeven coal prices to be greater than those shown in Table F-3. For example, if the cost of constructing coal and nuclear plants both rose by 10 per cent in real terms, maximum competitive coal prices at an operating rate of 5 000 hours per year could be roughly $7 and $14 per t.c.e., greater for high and low-sulphur coals respectively than the prices shown in Table F-3, and coal's competitive position vis-à-vis nuclear power would be significantly improved.

In a somewhat broader context, coal's competitiveness with nuclear power (which appears to be largely determined by local conditions) is not the main subject of this report. In regions where the comparative electricity generation costs in new coal and nuclear power stations are relatively close, the risks associated with future uncertainties could be minimized by supplying the additional generating capacity from both fuel sources.

The important conclusion to be drawn from this analysis is that both coal and nuclear power are likely to produce electricity at a lower cost than oil-fired plants in future years and in some respects, to the extent that there are obvious advantages associated with diversifying the sources of electric power, coal and nuclear power may be more complimentary than competitive.

COAL PRICE AND QUALITY DIFFERENTIALS

Tables F-1, F-2, and F-3 show that the additional capital, operation and maintenance and fuel costs of owning and operating flue gas desulphurization facilities can add a considerable amount to the cost of electricity generation in new coal-fired power stations. Likewise, the high cost of FGD facilities can have an important impact on the prices utilities are willing to pay for a particular quality of coal.

In general, for a given environmental limit on emissions of sulphur dioxide (SO_2) a utility would choose to buy a higher-sulphur coal that required scrubbing only if its delivered price were sufficiently below the delivered price of low-sulphur coals (that do not require scrubbing) to offset the cost of cleaning the flue gas. Table F-1 shows that the cost of electricity generation in a coal plant burning low-sulphur coal without scrubbers (FGD) is 6.2 mills per kwh less than the cost of electricity produced in a plant with a scrubber capable of cleaning 100 per cent of the flue gas. In this case, utilities could afford to pay a premium on the delivered price of low-sulphur coals that do not require scrubbing of $0.60 per 10^6 BTU (or $16.60 per t.c.e.) over and above the delivered price of high-sulphur coals that require FGD and still generate electricity at the same cost[9].

The actual premiums/discounts on the delivered prices of high and low-sulphur coals that utilities would be willing to pay will depend on the sulphur content of the coal and on the proportion of the flue gas that must be cleaned to obtain a given limit on SO_2 millions. Table F-4 shows, for example, the proportion of the flue gas requiring desulphurization to achieve an SO_2 limit of 1 lb. SO_2 for each million BTUs burned[10], for 10 coal types with a high and low BTU content and sulphur contents (by weight) varying from 0.5 per cent to 4.5 per cent. Taking into account variations in the heat rate requirements and the cost of flue gas desulphurization, Table F-4 also shows the required discounts on the delivered prices of the various coals to offset the cost of the desulphurization required to meet a 1 lb. SO_2 standard.

The Table F-4 shows that high BTU/high sulphur (4.5 per cent) coal would be an economic choice for a utility only if its delivered price were $13.80 per t.c.e. below the delivered price of a low-sulphur coal that could be burned without scrubbing. The maximum competitive or breakeven prices for most of the coals likely to be available in the OECD area will therefore fall somewhare between the maximum competitive coal prices for high and low-sulphur coal shown in Tables F-2 and F-3, depending on sulphur content and the proportion of the flue gas requiring scrubbing to meet the relevant SO_2 limit in different areas[11].

The discounts/premiums on high or low-sulphur coals will only be relevant in regions where several different types of coal are competing with one another in a given market. In addition, the design of the environmental regulations will have a major impact on coal price/quality differentials. The analysis presented above is based on the cost of scrubbing required to meet a specific, numerical limit on SO_2 emissions. However, if the environmental regulations specify that all coal plants utilize the best available control technology in all circumstances (or that all coal plants remove a given percentage of the sulphur dioxides from the flue-gas) regardless of the quality of

---

9.   Alternatively, utilities would only be willing to buy high-sulphur coals that require 100% FGD if their delivered prices were sufficiently discounted below the price of low-sulphur coal to offset the cost of scrubbing.

10.   0.45 kg SO_2 per 2.5×10^5 kcal. As a point of reference the federal standard in the United States is currently 1.2 lb SO_2 per 10^6 BTU.

11.   The differentials presented in Table F-4 are based on the cost of lime/limestone flue gas desulphurization only. Any technological developments that lower the cost of desulphurization will tend to narrow these differentials.

the coal burned, the cost of desulphurization would apply more equally to all coal types, and the differentials on the delivered prices of heat-equivalent units of coal related to sulphur contents would greatly diminish or disappear.

Table F-4   **Discounts Required on the Delivered Prices of Selected Coal Types to Offset the Cost of Flue Gas Desulphurization**

Coal Type	Per cent Sulphur to Weight	Portion of the Flue Gas Requiring Scrubbing[3]	Required Discounts[1] (In 1976 $ U.S.)		
			mills/ kwh[2]	$ per 10^6 Btu	$ per tce
	(percentage)	(percentage)			
High BTU Coal[4]					
1	0.5	0	—	—	—
2	1.5	18	1.1	0.11	3.17
3	2.5	55	3.3	0.34	9.57
4	3.5	71	4.3	0.44	12.28
5	4.5	80	4.9	0.50	13.78
Low BTU Coal[5]					
1	0.5	0	—	—	—
2	1.5	49	3.0	0.31	8.55
3	2.5	74	4.5	0.46	12.77
4	3.5	84	5.1	0.52	14.44
5	4.5	90	5.5	0.56	15.44

1.   Below the delivered price of a low-sulphur coal that could be burned without FGD.
2.   At 5 500 hours per year.
3.   The portion of the flue gas that has to be scrubbed to reach an SO_2 limit of 1 lb. SO_2 per 10^6 BTU (0.45 kg per 2.5 × 10^3 kcal).
4.   High BTU coal: 12 000 BTU/lb. or 6 665 kcal/kg.
5.   Low BTU coal: 8 000 BTU/lb. or 4 443 kcal/kg.

COAL COSTS AND PRICES IN THE FUTURE

General Assessment

The coal industry, due to its relatively large untapped resource base, is usually characterized as a constant cost industry, and coal supply curves are generally expected to remain relatively flat in real terms over the foreseeable range of coal demands likely to persist throughout the rest of this century. The issue, however, is not without debate. For example, some observers believe that coal prices are primarily a function of wolrd oil prices, that coal production and transportation costs will rise more rapidly as output increases, or that increased market power of the various agents in the different segments of the coal supply chain will allow a wider margin between coal costs and coal prices in the future.

Even though we have not attempted any systematic forecast of future coal costs and prices for the main producing and consuming areas, our analysis of these issues has led us to conclude that, in the long-term, coal prices in real terms are likely to rise less rapidly than oil-prices for a number of reasons. Coal prices did rise substantially after the oil price rise in 1973, but in all instances, the difference between coal and oil prices on a heat content basis did widen substantially. This was even true in the United States where the prices of oil and natural gas were artificially controlled below world market levels while the uncontrolled price of coal was free to seek whatever level it might attain.

In the absence of the limitations on the use of imported steam coal that currently exist in some countries, there is considerable competition in the world coal industry.

Competitive forces in the international coal markets are enhanced by the widespread nature of the abundant reserve base, a lack of corporate concentration, few institutional constraints to entry (with the possible exception of access to capital), and competition from the suppliers of other forms of energy. Furthermore, the promised development of new coal burning, cleaning and blending technologies will allow greater use of a wider variety of coal types in most facilities in the future, leading to increased competition among coal supplies. The many alternatives available for satisfying future coal demands may interact to create a situation where changes in coal mining or transport costs in a particular region produce significant changes in the mix of coals demanded from various supply sources, but relatively small changes in the price of coal delivered to consumers. Government policies to promote freer trade and investment in steam coal could also enhance and insure greater competition in world coal trade.

Nevertheless, in spite of the abundance of coal reserves, more stringent mine safety and health regulations and environmental controls affecting land reclamation may raise overall coal production costs. These considerations, however, which may shift coal supply curves upward by the cost of compliance with the regulations once and for all, are not likely to be the source of constant upward pressure on coal costs and prices.

There is also considerable uncertainty over the future supply of labour and labour costs in the coal mining industry. However, even if hourly wage costs in coal mining rise slightly more than average hourly wages in the economy as a whole, the overall effect on units labour costs is likely to be moderated by increases in labour productivity result from:

  i)   the gradual introduction of new mining technologies,
  ii) the opening of new, larger and more productive underground coal mines, and
  iii) the expected shift to large-scale surface mining in major producing countries.

Nevertheless, there may be temporary shortages of mine labour in some regions where coal production is expected to increase rapidly, but there are limits to the differential rate of increase in miner's wages over and above wage increases for the economy as a whole necessary to attract the required labour supply. Government manpower policies to promote training and relocation, and government policies designed to promote additional competition and freer trade in coal markets could also alleviate wage pressures.

The discussion of the world coal reserve base presented in Chapter D indicates that worldwide coal mining costs should not come under pressure from gradual depletion of low cost coal reserves and a shift to less productive marginal mines for a long time into the future. In the long-run, however, coal mining will undoubtedly be affected by the same forces that affect other extractive industries: coal production costs will rise as less productive marginal mines are developed, deeper and thinner seams are exploited, and as over-burden ratios rise in strip-mining areas. Higher production costs will affect coal's competitive position vis-à-vis the cost of alternative energy supplies. Nevertheless, higher cost marginal coal mines will only be developed if there is sufficient demand for coal at prices that will cover costs, and there will only be sufficient demand for coal if coal is a competitive source of energy.

At present, coal and oil compete directly primarily in the under-boiler fuel markets in industry and electricity generation. In the longer-term, however, as oil supplies become increasingly more expensive, coal and oil may be consumed in entirely different markets which will tend to weaken any direct link between coal and oil prices that may currently exist. At that time, only the products from coal conversion processes (liquids, gas, and electricity) will be direct competitors with oil, and the differential between coal and oil prices on a heat equivalent basis will have to be sufficiently large to allow the high cost of coal conversion and the production of competitive products.

In the foreseeable future, therefore, coal prices are likely to be determined less by competition between coal and oil directly, and more by intrafuel competition among coal suppliers and by competition between coal and nuclear power in the electricity generation sector. Rising oil prices may affect the price of coal by increasing the demand for coal and by increasing coal production and transportation costs, but oil prices are unlikely to be the main determinant of coal prices as some believe. On the contrary, a widespread shift to lower-cost coal in those uses where coal can readily substitute for oil may limit the rate of oil price increases in the future.

Steam Coal Prices in OECD Regions.

The future price of steam coal is likely to be dominated by production and transportation costs and, in many regions of the OECD (especially coal-importing countries) transportation costs could amount to more than 50 per cent of the total delivered price. Actual coal prices, especially spot market prices, will also be influenced by short-term fluctuations in supply and demand and they are extremely difficult to estimate for a given point of time in the future. Estimates of the cost of steam coal delivered to different regions of the OECD, however, are somewhat less uncertain than estimates of price. Delivered coal cost estimates include all direct and indirect cost elements, including capital charges, but no allowance for abnormally large swings in the profit margins of the coal and transportation industries for cyclical or other reasons.

In the paragraphs that follow, the maximum competitive prices for high and low-sulphur steam coal shown in Tables F-2 and F-3 are compared with the expected future supply prices for steam coal in the major OECD countries and regions to determine the competitive position of steam coal vis-à-vis oil and nuclear power in electricity generation. The analysis of future supply prices utilizes the transportation cost estimates and the coal production cost estimates developed in other sections of this report, as well as estimates derived from other sources.

In the *United States*, a large number of independent studies have produced estimates of coal prices at the mine mouth and delivered coal prices to electric utilities in the different regions of the country to the year 2000. In general, the majority of the studies conclude that real coal prices will escalate some 1-2 per cent p.a. from now to the year 2000 for the nation as a whole, with considerable differences in the escalation of real prices at the mine mouth in the various coal producing regions[12].

The national average price of coal delivered to utilities in the US is currently around $0.90 per 10^6 BTU in 1976 dollars ($25 per t.c.e.), with a regional high of $1.30 in New England ($36 per t.c.e.), and a low of $0.42 in the Mountain states of the West ($11.65 per t.c.e.). These prices compare very favourably with the national average price of fuel oil delivered to utilities which is currently $2.25 per 10^6 BTU, ranging from a low of $1.80 in the states bordering the Gulf of Mexico to a high near $3.00 in the West North Central states. Furthermore, even if the national average price of coal delivered to utilities rises by 2 per cent per annum in real terms to the year 2000, the national average price of steam coal would only rise to $1.39 per million BTU or $39 per t.c.e. (in 1976 dollars), a price that is considerably below even

---

12. It should be noted, however, that the oil price assumption included in most of these studies is considerably greater than the 2½ per cent p.a. increase included in this report.

today's oil price. Moreover, if oil prices rise, which is most likely considering that oil prices in the United States are currently regulated below world market levels, the competitive advantage of coal over oil will be even greater.

A number of other studies generally conclude that coal will clearly be competitive with nuclear electricity generation in new baseload power plants in the low-cost coal mining states of the West and in the West South Central region. Nuclear power, on the other hand, appears to have a competitive edge over coal in baseload operations in New England, and in parts of the Middle and South Atlantic regions, and more generally in densely populated areas where environmental regulations effectively prohibit the burning of coal. In other regions the choice between coal and nuclear power is less definitive and likely to be determined by local conditions. In intermediate and peak load operations, however, coal will generally have a competitive advantage over nuclear power. Given the rather small differences between the cost of coal and nuclear power in most regions of the United States and the large competitive advantage of both coal and nuclear power over oil-fired electricity generation, a widespread shift away from oil in the electricity sector as proposed in the current administration's National Energy Plan is clearly justified on the basis of relative costs. However, if the present delays in the implementation of nuclear power result in a delayed shift away from or an increase in oil consumption, the prices of electricity charged to consumers could rise substantially, oil imports could be higher, and the balance of payments could suffer further deterioration.

Canada.   Estimates of the delivered cost of steam coal to new power stations range from $.30 per million BTU for mine-mouth power stations burning low rank coals in the Western Provinces, to $1.25-$1.50 per million BTU in the Maritime Provinces, and $1.40 and $1.75 per million BTU for imported coal from the United States and Western Canadian bituminous coals, respectively, delivered to utilities in Ontario. These prices are clearly below the maximum competitive prices for steam coal to be competitive with oil shown in Table F-2.

While the analysis of nuclear power costs presented in Table F-1 is less applicable to Canadian licensing requirements and the CANDU-pressurized heavy water reactor system built in Canada, other estimates have shown that coal may have a competitive advantage over nuclear power only in the lowest-cost coal mining regions of the Western Provinces.

Japan.   Stringent environmental regulations are likely to require that all electric power stations utilize highly efficient flue gas desulphurization technologies regardless of the sulphur content of the coal burned. In addition, strict limits on nitrogen oxide emissions are likely to add an additional $20 to $40 per kw to the cost of constructing new power stations in Japan, although the technology is not yet fully developed for coal-fired boilers. Furthermore, coal's competitive position may suffer from siting problems and high land prices that will raise the costs of coal handling, distribution and burning facilities, and the cost of sludge and ash disposal systems. Nevertheless, as oil prices rise, imported coal is likely to enjoy an increasing competitive advantage over oil in electricity generation.

The Industrial Research Institute of Japan has recently completed a study of future imported steam coal prices in Japan from Austrialia, USA, Canada, Mozambique, India, Indonesia, Colombia, South Africa and China. These estimates, which have been converted to delivered prices per ton of coal equivalent, are presented in Table F-5.

Based on the analysis of maximum competitive coal prices presented earlier, imported coal burned in a new power station with flue gas desulphurization facilities will be competitive with oil-based electricity generation in baseload operation at delivered coal prices equal to or less than $71 and $73 per t.c.e. at operating rates of 5 000 and

7 000 hours per year respectively[13]. These prices would fall by roughly $ 1.50 to $ 2.50 per t.c.e. when an additional $ 25 to $ 40 per kw is added to the capital cost of coal-fired power stations shown in Table F-1 to offset the higher land cost and environmental control regulations applicable in Japan.

Even so, all the estimates of the future price of imported steam coal in Japan in Table F-5 are well below the maximum competitive price utilities could afford to pay for steam coal and still generate electricity at the same cost as oil power stations. Furthermore, even at today's oil price, imported steam coal from a number of supply regions shown in Table F-5 would be competitive with oil in electric power stations, especially if oil plants also require expensive FGD facilities.

The high transportation cost component of the delivered price of steam coal in Japan and the need to burn coal in an environmentally acceptable manner raises doubts about the competitive position of imported steam coal vis-à-vis nuclear power in future years[14]. Nevertheless, strict licensing requirements and siting problems have already adversely affected the cost of nuclear power to the point where its competitiveness with imported coal in the future is at least open to question. However, since both coal and nuclear power are likely to produce electricity at a lower cost than oil-fired power stations in the future, Japan could gain a measure of diversity and security of supply and minimize costs by using additional coal rather than oil for any future shortfalls in nuclear implementation plans.

In *Western Europe*, all the imported steam coal and most of the domestically produced steam coal (with the possible exception of the highest-cost supplies in Germany) available today is competitive with oil in electricity generation in spite of the depressed state of heavy fuel oil markets. In addition, natural gas, which previously competed with coal in electric power generation, is gradually being phased out of the electricity sector.

Coal will be competitive with nuclear power in baseload operations in Western Europe if low-sulphur coal (that can be burned without expensive FGD facilities) can be delivered to utilities at prices equal to or less than $ 32-$ 36 per t.c.e. (or $ 46-$ 52 per t.o.e.). Table F-5, which presents estimates of the future prices of imported steam coal delivered to major ports in Western Europe shows that the steam coals imported from Poland and South Africa could possibly be competitive with nuclear power in baseload operations in power stations near coastal areas or with easy access to water transport systems.

From an economic viewpoint, therefore, it would be most advantageous for European power stations to substitute imported steam coal rather than oil for any shortfall in their nuclear development programmes. However, the decision of an individual utility will generally be based on consideration of all the options available at the time of decision. A given power station may be built because it is the most economic of all the plants available, or because it is the most economic choice given the constraints of lead times. In the absence of more coal stations in some Western European countries, a shortfall or delay in nuclear power programmes may be met by upgrading an existing oil plant from a lower to a higher operating rate, or by building new oil plants[15]. If more coal plants were available or being planned, however, the overall costs of a nuclear shortfall could be minimized by accelerated completion of coal plants under construction or by possible upgrading of existing coal plants to take

---

13. $ 101 to $ 104 per t.o.e. respectively.

14. The low operating rate experience in Japan's nuclear power plants (40 per cent) in the recent past is currently raising the cost of the nuclear kwh, but this is judged to be a temporary phenomenon.

15. Coal plants, because of their competitive advantage, are likely to be already operating at higher rates than oil plants.

Table F-5   **Estimated Costs of Imported Steam Coals Delivered to Western Europe
and Japan in the Mid-1980's[1]**

1976 $ U.S. per t.c.e.

Supply Region	Quality (Sulphur Content)	Estimated C.I.F. Cost per t.c.e.	
		Western Europe	Japan
United States			
Eastern	L.S.	$51 - $55	$62 - $64
	H.S.	$38 - $40	$49 - $50
Western	L.S.	$48 - $54	$32 - $36
	H.S. - H.M.S.	$38 - $51	$36 - $43
Canada			
Eastern	H.S.	$40 - $45	
Western	L.S.	$46 - $50	$42 - $46
South Africa	L.S. - M.S.	$34 - $42	$35 - $43
Australia	L.S.	$44 - $47	$34 - $46
Poland	L.S. - M.S.	$34 - $42	
USSR			$37
China			$37 - $40
Indonesia			$37 - $48
Colombia			$35 - $41
Mozambique			$40

1. First year prices of long-term contracts.
L.S.: low-sulphur
M.S.: medium-sulphur
H.S.: high-sulphur.

up the slack. Since the difference between imported coal and nuclear generation costs is considerably less than the difference between oil and nuclear, substitution of coal rather than oil in the event of a nuclear shortfall would minimize the overall impact of a nuclear delay on electricity generation costs as well as the balance of payments.

SUMMARY-COAL'S COMPETITIVENESS IN ELECTRICITY GENERATION

Table F-6 present regional estimates of electricity generation costs in new nuclear, oil and coal-fired power stations in Japan, Western Europe and the United States. The cost estimates for nuclear and oil-fired electricity generation and for capital and operation of coal plants with 100 per cent FGD are taken directly from Table F-1[16].

The coal price estimate for the United States is based on the most recent forecast of steam coal prices delivered to utilities in 1985 by the Department of Energy, and it includes the impact of the United Mine Workers recent wage settlement and real coal price escalation of 1 per cent per year beyond 1985.

The coal price estimates for Western Europe and Japan shown in Table F-6 are based on a simple average of the estimates of imported coal shown in Table F-5 for each region, and they include an additional $4.50 per t.c.e. to cover terminal port charges and final delivery costs as well as real coal price escalation (including port charges and final delivery costs) of 1 per cent per year after 1985[17]. The simple

16.   Except in the case of Japan where $25 per kW have been added to the cost of the coal plant to allow for the costs of additional environmental protection devices and higher land costs.

17.   The $4.50 includes $1.50 per tce for terminal unloading costs and $3.00 per tce for tranship-ment by barge or rail, (see Chapter E).

Table F-6 Regional Estimates of Electricity Generation Costs in New Baseload, Nuclear, Oil and Coal-Fired Power Stations[1]

In 1976 dollars, Average Costs per kwh during the first 20 years of operation for new plants commissioning in 1986

Average Costs per kwh over the first 20 years ($ mills)	All Regions (from Table F-1)			Japan	Western Europe		United States
	Nuclear	Fuel Oil		Imported Coal[3]	Imported Coal[3]		Domestic Coal
	PWR 2 x 1100 MW	2 x 600 MW		2 x 600 MW	2 x 600 MW		2 x 600 MW
		Low-Sulphur HFO Conventional	High-Sulphur HFO with 100% FGD	With 100% FGD	Without FGD	With 100% FGD	With 100% FGD
Capital Cost	14.9	7.5	9.6	12.9	9.6	12.4	12.4
Operation and Maintenance Costs	2.4	2.0	4.2	5.1	2.2	5.1	5.1
Fuel Cost	6.5	31.0	29.0	17.8	18.5	18.5	12.8
Total Average Cost per kwh at 5 500 h/a	23.8	40.5	42.8	35.8	30.3	36.0	30.3
at 6 500 h/a	21.6	39.3	41.3	33.8	28.8	34.1	28.4
at 5 000 h/a	25.3	41.2	43.8	37.1	31.3	37.2	31.5
at 4 500 h/a	27.2	42.1	44.9	38.7	32.4	38.7	33.0
Cost of Construction ($/kW)	$700	$350	$450	$605[2]	$450	$580	$580
Average Fuel Cost for the period 1986-2006 :							
$ per toe (10^7 kcal)	$25.79	$132.00	$118.00	$71.30	$74.00	$74.00	$51.40
$ per toe (0.7×10^7 kcal)	$18.05	$92.40	$82.60	$49.90	$51.80	$51.80	$35.98
$ per 10^6 BTU	$0.65	$3.33	$2.97	$1.80	$1.87	$1.87	$1.30
Memorandum Item :							
Total Cost per kwh							
at 5 500 h/a in : 1986		34.8	36.4	34.1	28.5	34.2	29.1
($ mills) 1990		37.4	38.8	34.8	29.2	34.9	29.6
2000		45.2	45.8	36.5	31.0	36.7	30.8

1. The Capital cost estimates, plant heat rates, and oil price assumptions are the same as those expressed in Table F-1
2. Includes $ 25/kW additional investment in pollution control devices. See text.
3. The coal prices shown in this table refer to estimates of the price of steam coal delivered to utilities in 1985, and they include real coal price escalation of 1 % p.a. after 1985. In the case of Western Europe and Japan, the coal prices shown are based on an unweighted average of the estimated C.I.F. imported coal prices shown in table F-5, and they include an additional $ 4.50 per tce (which also escalates at 1% p.a. after 1985) to cover terminal port charges and final delivery costs. See text.

h/a : hours per year
FGD : flue gas desulphurization

average of the imported coal prices shown in Table F-5 will overstate the actual weighted average imported coal price since it is likely that higher volumes will be attached to the lower-cost coal supplies.

Nevertheless, Table F-6 shows that the cost of electricity generation in new coal-fired stations is likely to enjoy a considerable advantage over the cost of electricity generation in new oil-fired power stations in most segments of the load curve over the life of the power station. Furthermore, the differences become even more pronounced with the passage of time: the differences between the cost of coal and oil-fired electricity generation rise from 7.3 mills per kwh, 2.2 mills per kwh and 2.3 mills per kwh in the United States, Western Europe and Japan, respectively in 1985, to 15 mills per kwh, 9.1 mills per kwh, and 9.3 mills per kwh, respectively, in the year 2000.

On the basis of the preceding analysis, therefore, it appears that coal is or will be considerably less expensive than oil for electricity generation in many OECD countries.

POSSIBILITIES FOR COAL SUBSTITUTION IN OTHER SECTORS

Table F-7 summarizes the major changes in coal consumption in the different sectors and regions of the OECD between 1960 and 1976. Coal consumption in the residential/commercial sector between these years fell by 80 Mtoe, followed by declines of 47 Mtoe in the industrial sector excluding iron and steel, 26 Mtoe in transportation, and 15 Mtoe in gas manufacture. The prospects for reversing these trends during the rest of this century are discussed below.

Table F-7　**Consumption of Solid Fuels by Sector in 1960 and 1976**

In Mtoe

	OECD		North America		OECD Europe		Japan	
	1960	1976	1960	1976	1960	1976	1960	1976
Gas Manufacture	19.2	4.6	0.1	0.2	13.7	2.4	4.7	2.0
Energy Sector Use and Losses	46.0	43.8	16.1	15.4	26.7	17.6	1.3	8.4
Total Industry	221.7	182.2	80.0	74.0	115.7	65.2	20.5	35.0
Iron and Steel	106.9	114.7	42.3	33.6	56.5	45.2	6.6	32.6
Others	114.8	67.5	37.7	40.4	59.2	20.0	13.9	2.4
Transportation	26.7	1.0	1.7	0.1	19.5	0.9	4.0	*
Residential/Commercial	129.3	49.2	27.6	6.5	96.3	37.9	4.7	3.7
Electricity Generation	227.8	425.5	117.9	268.8	88.4	130.0	14.3	8.8
Total Coal **	675.4	707.8	243.4	366.9	364.8	253.0	49.6	58.2

* Less than 0.1.　** Total includes statistical differences.

Residential/commercial Sector

In the near-term, coal may find its way into the residential/commercial sector mainly in its transformed condition through the medium of electricity. Space heating/cooling, hot water, and cooking account for most of the energy consumed in the residential/commercial sector, and electricity is a convenient and readily available substitute for oil and gas for these purposes. In 1976, however, electricity accounted for only 20 per cent of final energy consumption, while oil and gas accounted for 45 per cent and 30 per cent, respectively. In most instances, however, the cost of converting existing units from oil and gas to electricity is prohibitive, and most of the growth in electricity consumption will occur in new buildings or through normal replacement.

The relative competitiveness of electricity vis-à-vis oil and gas in new buildings will be the main factor determining electricity's share in the final energy consumption of this sector. Delivered electricity prices per BTU of useful energy are generally higher than oil and gas prices in the residential/commercial market, but this is offset to a certain degree by electricity's higher end-use efficiency, cleanliness, and convenience, and by the lower capital cost and smaller space requirements of electrical implements. In addition, a large part of the heat requirements of this sector could be supplied by cheaper, baseload power plants during off-peak hours. The current low price of natural gas and the prospect of ample supplies in some regions in the immediate future, however, could limit the rise of electricity's share over the near-term. The substitution of coal for natural gas in other sectors, however, may allow greater substitution of the freed-up gas for oil directly in the residential/commercial sector.

In the longer-term, the prospects for coal in the residential/commercial sector could be enhanced by successful commercialization of fluidized bed combustion and more interaction between energy and urban and regional development planners. Cleaner coal burning technologies could allow greater use of coal-fired boilers in large apartment complexes, office buildings, commercial centres, schools and other institutions through a number of variations and extensions of district heating schemes and the combined production of heat and power from central plants serving the needs of new communities.

Industrial Sector

Coal consumption in the industrial sector, excluding the iron and steel industry, fell considerably between 1960 and 1976. Between these years, however, many newer coal burning facilities were also converted to oil and gas. The immediate reconversion possibilities in major fuel burning industrial plants (MFB's) previously burning coal is difficult to establish. In some instances, the coal burning and handling equipment may have been dismantled or completely replaced, and environmental regulations and space requirements probably limit the potential for reconversion. Few surveys of the fuel-burning capabilities of industrial MFB's are available.

While the potential for reconversion to coal in existing facilities may be small, the potential for conversion and coal substitution in the industrial sector over the long-run is considerable. Industrial energy consumption accounts for roughly 30 per cent of the OECD's primary energy requirements, and 29 per cent and 36 per cent, respectively, of the area's total consumption of oil and natural gas. Furthermore, oil and gas consumption currently account for roughly 65 per cent of industry's total energy demand. Apart from feedstock requirements, the captive markets for electricity, and processes requiring extremely clean fuels or precise temperature control, coal could potentially supply a large portion of the industrial sector's direct heat and process steam requirements currently supplied by oil and gas.

Fuel consumption in the industrial sector, excluding electricity and feed-stocks, is split fairly equally between boiler fuel for captive electricity and process steam generation and direct heat in large kilns, heaters, ovens, and dryers. The process steam or heat requirements of the chemical, petroleum refining, paper, building materials, and food processing industries appear to be good candidates for coal substitution in the long run. However, for practical reasons, only the largest establishments may have the capability of converting to coal. The possibilities for coal substitution in smaller establishments and in non-boiler (direct heat) applications outside of the cement and building materials industry remains ill-defined[18].

In the near-term, however, there are a considerable number of technical, economic and environmental constraints limiting the possibilities for coal substitution in the industrial sector. The conversion of coal-fired equipment to oil or gas is much easier and cheaper than conversion of oil and gas equipment to coal. In many instances, conversion to coal is impractical since it would require the use of different boilers, heat exchangers, and refractory surfaces. Furthermore, conversion to coal requires allocation of space for the coal pile, installation of coal handling equipment, provisions for ash collection and disposal, dust control, and possibly flue gas desulphurization equipment. The design and location of many industrial establishments may have been strongly influenced by the use of oil or gas in the past, and space may not be available to allow conversion to coal. Furthermore, environmental regulations may effectively prohibit coal consumption in densely populated areas or regions where air quality problems already exist, or strict environmental standards may involve higher costs in terms of a premium for clean fuel prices or in terms of higher pollution control costs for all energy facilities. There are also limits to the existing distribution system's ability to supply smaller-scale industrial demands in diverse locations at reasonable cost.

The substitution possibilities and ease of conversion between oil and gas currently provides industry with a measure of flexibility and additional security without the high capital expenditures required for conversion to coal. In addition, the relatively low price of natural gas in some existing long-term contracts and the regulated price of natural gas in certain markets also reduces the incentive for coal substitution in the near-term. Government energy policies to reduce oil and gas consumption in electric power plants may also be enhancing the industrial sector's perception of the longer-term supply prospects for these fuels and reducing the perceived risk of oil and gas curtailments.

Nevertheless, coal is currently a competitive fuel in largescale industrial boilers for generation of electricity and process steam, and in certain direct heat applications. The high cost of converting existing facilities, however, may limit the near-term increase in coal consumption to new facilities designed to burn coal from the start, and to the normal replacement market as the existing stock of capital equipment turns over (large industrial boilers must generally be replaced every 20-30 years). Furthermore, as oil and gas prices rise, the possibilities for coal substitution will improve. It is difficult, however, to estimate the cost of industrial conversion to coal because of the wide variety of local circumstances, economies of scale, and the site-specific nature of the process heat or steam requirements in individual establishments. Installation of coal-burning equipment in industrial MFBs, however, may cost 2-5 times as much as oil and gas equipment. Access to low-sulphur coal supplies could also be an important factor since the high cost of existing flue gas desulphurization equipment limits installation of these devices to only the largest industrial plants.

In the longer-term, successful commercialization of fluidized bed combustion technology, and improvements in coal cleaning, blending and distribution systems will limit some of the current environmental constraints, space limitations, and other objections to industrial coal substitution. At the present time, many industries capable of converting to coal have indicated they are delaying conversion plans for a few years in

---

18.   The major problems of coal in non-boiler uses appear to be (i) product contamination by coal ash, dust, sulphur and trace minerals, (ii) the breakdown of refractory surfaces, (iii) the difficulty of precise temperature control due to the variability of coal supplies and coal combustion, and (iv) corrosion that reduces the efficiency of heat exchangers.

order to take full advantage of the promises and prospects offered by fluidized bed combustion.

In the longer-term, several other promising developments could enhance the possibilities for industrial coal consumption. Improved interaction between energy and regional development planners could foster the development of industrial parks where the energy requirements of many smaller firms (who might find conversion to coal impractical on their own) could be supplied by central coal plants capable of taking full advantage of the scale economies and higher efficiencies offered by cogeneration. Gasification of coal into clean low and medium BTU industrial fuels in centralized plants serving local industrial needs will also offer higher efficiencies and more possibilities for coal substitution by allowing combined-cycle steam and electricity generation and the use of coal-gas in a wider range of industrial applications.

OTHER FAVORABLE ECONOMIC CONSEQUENCES OF GREATER COAL USAGE

In addition to the lower long-term costs of coal-fired plants compared to oil-fired plants, some additional economic gains from substituting coal for oil are briefly cited below:

- A large part of the coal used in future years will come from indigenous coal reserves within the OECD area, providing employment opportunities in the coal mining and related equipment and transportation industries.
- Potential coal importing countries, with little or no indigenous coal resources, will realize substantial foreign exchange savings by importing lower-cost coal in place of higher cost oil for use in electricity generation or other large-scale industrial applications where coal can be readily substituted for oil.
- Increased utilization of steam coal in the OECD will set a favourable example for the non-oil developing countries, and promote accelerated exploration and development of their own resources. Increased coal utilization and coal exports from developing countries will have a favourable impact on their external payments position that could enhance world growth.
- Increased utilization of coal in the near future may speed the development of new technologies that would allow even greater substitution of coal-based liquid and gas fuels for oil in the period beyond 1990, and higher utilization of the existing infrastructure for liquid and gas fuel distribution.
- The substitution of coal for natural gas in certain sectors able to use coal will allow substitution of the freed-up natural gas supplies for oil in other sectors where the immediate possibilities for direct coal substitution are more limited.
- Greater coal substitution will enhance national security by allowing further diversification of energy supplies and reductions in oil-import dependency.
- Increased utilization of coal worldwide will reduce pressures on world oil reserves and retard the speed of oil price increases in the future.

G

CURRENT AND PROSPECTIVE COAL BURNING, MINING, CONVERSION AND TRANSPORT TECHNOLOGIES[1]

There exist at present a number of developing technologies for coal mining, transport and utilization which promise significant potential in the time frame to 2 000 as improvements in the use of coal for traditional purposes or as new or alternative energy sources throughout the world. The developing technologies which offer potential for increased usage of coal lie in the following areas — fluidized bed combustion, low and high BTU gas and liquids from coal, power conversion such as fuel cell, and magneto hydrodynamic (MHD) systems directly utilizing coal derived fuels. Additional technological advances that appear promising are in coal mining, pipeline transport and sea cargo handling facilities. Implementation and lead times of those technologies will vary widely in displacing current technology or fuels.

The following coal technology assessment presents a brief description of operation of each concept or process, its uniquely attractive features and its state of the art. Also included is a description of technical or other problems which impede commercial acceptance and an estimate of how long it may take these technologies to develop to the point of making significant commercial impact.

Where possible, estimates of capital costs and fuel or power costs associated with commercial application are presented, although these costs are often speculative and can be further refined only with greater development of the concept.

MINING TECHNOLOGY

Objectives

The introduction of advanced coal mining techniques could serve to increase productivity of the mining industry, increase the recovery of coal from the mined areas, and help to maintain or decrease the cost of coal. This must be accomplished in a manner that would provide for the safety of the miners and adequate environmental control.

State-of-the-Art

Deep Mining

Narrowing profit margins over the last two decades have dictated a trend in deep mining toward fewer but larger collieries with higher capacity, heavily mechanized, capital intensive producing units. In the most common European extraction system — longwall working — these producing units may consume 1-2 megawatts and produce

1. This chapter was written with the assistance of the staff of the NCB (IEA Services) Ltd. in London.

coal at up to 10 000 tons per day from the integrated complex of shearer or plough, armoured face conveyor, and powered roof supports. In the alternative extraction system – room and pillar mining – extensive R & D is being undertaken to match continuous haulage systems to the proven high capacity of the continuous mining machine (12-15 tons per minute).

This productive capacity at the face is now being matched by the introduction of sophisticated computer control systems both for effective integration of the production systems, and the better protection of both men and machines against the mine environment. Other extensive R & D efforts attempt to improve the effectiveness and specific cost of roadway and shaft drivage by automated systems.

However, rapid extraction rates will inevitably lead to increased depths of deep mining bringing new problems of heat and strata control. It is against these new constraints that much current R & D is aimed so that technology can be ready to meet any deterioration in natural conditions.

Surface Mining

For surface mining, the main changes taking place are those of scale rather than technique, although exceptional equipment capable of transferring up to 10 000 tons of earth per hour is working in some open pits, and new mines are planned to produce up to 20 million tons per annum of coal. The advantages of scale are not only those of cost reduction, but also in giving surface mining an increased ability to reach deeper seams.

Surface mining methods can be improved by the integration of mining and reclamation methods, the automation of the process to improve the utilization of equipment, and the optimization of the mining equipment itself. Current trends are toward the use of drag lines and the introduction of hydraulic excavators. With the requirements imposed by reclamation, a major improvement in the rate of production such as doubling, is not likely.

Both mining types rely in common on effective prospecting and delineation of the proposed deposit, and on the effective preparation of the product for the consumer.

Coal mine geophysical prospecting has been able to take advantage of extensive oil well technology and refine this to the particular needs of the coal industry, reducing dramatically both cost and time needed to effectively delineate a mine property.

Cost/Time Considerations

The introduction of advanced underground mining techniques which involve development and testing followed by the manufacture of new equipment required considerable time, 10 years perhaps. The opening of new mines requires some 5-7 years. Consequently, a major change in the mix of underground mining methods may not be realized until 1990 or later at current rates of development.

The production of heavy surface mining equipment affects the rate at which such mining can be increased and present projections indicate a total output doubling by 1985-1990 at the earliest.

COAL CLEANING

Introduction

For decades coal has been cleaned by physical washing to remove dirt and so produce a saleable, more easily transportable product. Increasing attention is now being paid to coal cleaning for sulphur removal in order to produce coals that meet environmental standards. At present, the extent to which coal cleaning is used varies considerably from country to country. Only 27 per cent of combustion coal in the US is cleaned compared with the relatively high proportions in Europe e.g. 64 per cent in the UK.

Objectives

The primary objectives of coal cleaning are at present to remove dirt and ash, and sometimes pyritic sulphurs contained in the coal, and so produce coal of improved or uniform heating value and ash content. This leads to economic benefits in terms of lower transport costs and more reliable boiler operation. With the increased concern over SO_2 emissions in recent years, physical coal cleaning practices have been carefully studied to determine how to maximize the sulphur removal, particularly in the United States and Germany. There is considerable research underway to develop chemical cleaning methods for sulphur removal.

State-of-the-Art

Present coal cleaning methods involve crushing of the coal to liberate impurites followed by physical washing to effect ash and pyrite removal using the difference in specific gravities of coal and its impurities. The gravity separation methods are relatively simple and utilize conventional equipment such as jigs, concentrating tables, hydrocyclones, dense medium gravity separators and dense medium cyclones.

In newer, more advanced systems froth flotation methods may be used. Separation is effected by the selective adhesion of air bubbles depending on the difference in surface characteristics between coal and its impurities. The primary use today is to improve recovery of valuable metallurgical coal fines but these methods can also be used to enhance pyritic sulphur removal.

Another modification of existing coal cleaning methods to improve sulphur removal and to minimize coal loss during cleaning is to clean the coal at two specific gravity separations to redistribute sulphur into a clean low-pyritic sulphur coal product and a middlings product with higher sulphur. The middlings product may then be used where there is post combustion control technology (FGD) or may be further cleaned to reduce sulphur. Methods are being investigated for further cleaning of the middlings which include regrinding followed by froth flotation.

While the physical coal cleaning methods just discussed will only remove dirt and some pyritic sulphur, chemical cleaning methods under development increase the potential of removing virtually all pyritic sulphur and, depending upon the technique, 0 to 50 per cent of organic sulphur in coal. However, the processes may not remove ash and mining waste and may need to be combined with a physical cleaning process.

Cost/Time Considerations

The modification of physical cleaning methods to improve sulphur removal for example by separating coal containing pyritic sulphur into a clean product and middlings is possible today. It is expected that more sophisticated means of secondary processing of the middlings such as froth flotation will be in more widespread use by 1985. The costs are summarized in the table below. If the plant operates 3 000 hours per year, the annual costs for the base plant are in the range of $ 1.80-2.80 per ton of raw coal or $ 3.00-4.70 ton of washed coal. Incremental costs for further sulphur reduction would range $ 0.80-1.50 over the base cost.

Unless there is a breakthrough in chemical coal cleaning technology, it is likely that deriving gaseous or liquid fuels from coal will offer greater environmental benefits.

Table G-1   **Costs for New Coal Cleaning Plants**
Currency at December 1976 values

	Capital Costs (currency/ton of raw coal/hr)		Operating Costs (currency/ton of raw coal)	
	FRG	UK	FRG	UK
Base cost to produce coal of uniform heating value and ash content	30-40,000 DM ($ 12-16,000)	£ 15,000 ($ 25,000)	2.5-3 DM ($ 1.00-1.20)	70 p ($ 1.17)
Incremental cost over base to redistribute pyritic sulphur into product and middlings	+ 12 % (assumes dense medium cyclones at SG = 1.30-1.32)		+ 5-16 %	
Incremental cost over base for redistribution and reduction of sulphur by secondary processing of middlings	+ 25-32 %		+ 20-40 %	

In that case, the technology may never become commercial. Otherwise, it is not expected that these methods will be used commercially until well after 1985. The costs remain undetermined at this time.

FLUIDIZED BED COMBUSTION (FBC)

Introduction

The fluidized bed combustion concept found its technical beginnings in the 1920s and since that time, significant development of FBC technology has taken place and it is approaching the time at which it can be considered available for commercial application. The term " commercial" is used here to mean that a supplier firm is willing to supply a fluidized bed combustor with full commercial guarantees as to its performance.

Objectives

Fluidized combustion technology offers a superior method of coal combustion in both the industrial steam raising/heating and the power generation markets. The advantages are the ability of a fluidized combustion plant to handle a wide range of coal (in type, quality and size) and the ability (through the injection of dolomite or limestone into the bed) to significantly reduce sulphur emissions from the plant. Also, because of its lower combustion temperature, nitrogen oxide emissions are reduced. Superior heat transfer characteristics in a fluidized bed plant offer the potential to reduce the size of plant by comparison with traditional coal-fired plant, and this improvement can ease the substitution of oil for coal in some industrial/power station situations.

There are two types of fluidized bed combustors under development. While atmospheric fluidized combustors (AFB) offer the advantages listed above compared with pulverized fuel power generation, it is likely that smaller scale atmospheric fludized beds (e.g. up to 50 MW thermal) will ultimately find their greatest use in enabling much cleaner use of coal for industrial use. On the other hand, pressurized fluidized beds (PFB) operating at pressures up to 20 atmosphere, are specifically aimed at obtaining greater efficiency of power generation by using a combined cycle power generation system to achieve greater energy conversion efficiency (see Table G-2).

In this context, the term "combined cycle power generation system" refers to units in which the normal steam cycle for generating electricity (which is driven by steam raised in the pressurized fluidized bed combustor) comprises only a part of the overall power generation system. After leaving the combustor, the hot flue gases are scrubbed of particulates and then expanded through a gas turbine which produces additional power, thus giving an improved efficiency for this type of system.

Figure G-1   **Simplified fluidized bed combustor: schematic diagram**

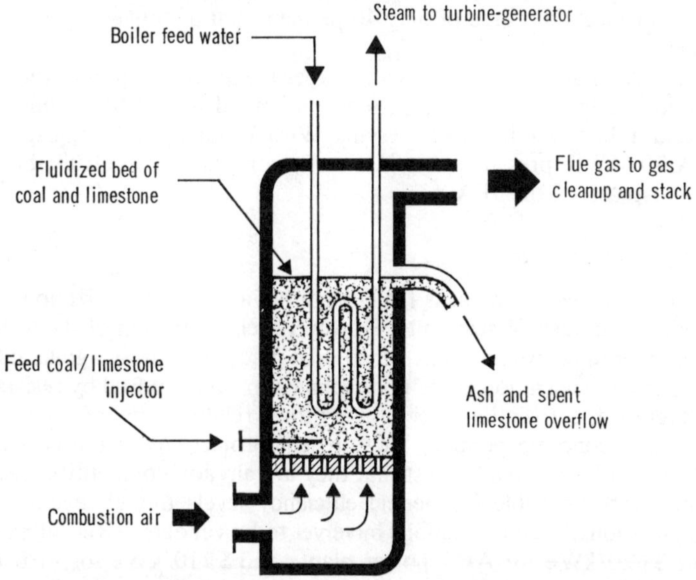

Table G-2   **Overall Efficiency Limits for Power Generation Systems Using Fluidized Bed Combustor/Boilers**

	Overall Coal-to-Electricity Efficiency
Atmospheric pressure systems	35-38 %
Pressurized systems	38-42 %
Conventional pulverized coal systems with flue gas desulphurization	31-34 %

In general, it is expected that the environmental impact of a coal-fired FBC system will be less than that of a corresponding conventional coal-fired system. Nitrogen oxide formation is significantly reduced and particulate matter is coarser and easier to collect. Water requirements for cooling and ash disposal are about the same as for conventional systems. The FBC system will have its greatest environmental benefit if an active bed of limestone or dolomite is used to absorb sulphur oxide (80-90 per cent removal). A limestone system will produce about the same volume of solid waste as a lime/limestone FGD system, but the waste is dry and may be easier to dispose of. However, it may be necessary to line the disposal area to prevent problems from rainfall leaching and other weathering problems. The waste disposal problem may be reduced but probably at greater cost with the application of systems for regeneration of the sorbent limestone. However, the development of these systems is some way off and costs need to be better determined.

State-of-the-Art

In an FBC, a bed of crushed coal is burned in air which is blown upwards at velocities sufficient to maintain fluidizing conditions. (See Figure G-1). Steam generation tubes are immersed in the fluidized bed where heat transfer rates are significantly higher than those found in conventional boilers. Because of this, installed heat transfer surface (and weight) is greatly reduced, with the result that atmospheric units (AFBs) may have a furnace plant area of only 40 per cent that of conventional systems (PFBs only 10 per cent).

There are numerous pilot-scale experiments throughout the world, but demonstration-scale, coal based experience is limited to a 8 MWe unit at Rendrew, Scotland and a 30 MWe unit at Rivesville, West Virginia, both atmospheric pressure systems. A 90 MWth pressurized unit is under construction at Grimethorpe, England under the auspices of the IEA.

Cost/Time Considerations

FBC process development is far enough advanced that utilization is technically feasible today in industrial size units. However, demonstration of the technology in a commercial operating situation has yet to take place. The most significant obstacle to achieving market penetration is the non-technical problem posed by resistance to leaving well-developed, conventional technology for a relatively unknown processing concept. From an economic point of view, recent projections of the cost of electricity produced with PFB systems suggest that they are already competitive with conventional systems at comparable (or better) efficiency levels though in absence of commercial applications, such projections have yet to be validated. Recent capital cost estimates are $620/kWe for AFB power plants and $710/kWe for PFB plants (mid-1975 dollars). These produce electricity at 3.2 and 3.5 cents/kWh respectively. A likely construction time for a FBC power plant is about 6 years.

ENVIRONMENTAL CONTROL TECHNOLOGIES

Introduction

Coal combustion produces a number of environmental burdens including:
1. Gaseous pollutants
2. Particulate emission (fly-ash)
3. Ash waste
4. Trace elements and organic chemicals

Thermal pollution, that is the rejection of a large amount of waste heat, is also associated with one form of coal combustion, the production of electricity with the conventional condensing turbine.

Objectives

All countries impose some level of environmental controls on coal combustion. The most general relate to ash control. The dumping of ash wastes is usually constrained by land use controls and fly-ash emission is invariably controlled above the residential use level by requiring the use of mechanized or electrostatic particulate removal apparatus (ESP). In addition to the control of ash, the control of smoke is often applied in urban areas by requiring the use of specially prepared smokeless solid fuel.

Many countries now require that emissions of sulphur oxides be controlled to some degree and Japan requires that emissions of nitrogen oxides be controlled. Countries such as Sweden, which at present do not burn coal, are likely to introduce control regulations, particularly over sulphur oxide emissions in introducing its use.

No country yet requires control over the emission of trace elements or organic chemicals. Some attention is being paid, however, to this matter, particularly emissions of uranium, mercury, lead and beryllium and carcinogens such as policyclic aromatic hydrocarbons.

State-of-the-Art

Solid waste disposal can be controlled either by improved dumping procedures or by using the ash in some further process. Both these are being pursued. Land reclamation of quarries and gravel pits is being undertaken with waste solids though suitable sites are limited. In certain countries, a high proportion (up to 60 per cent) of solid waste from power stations is used as aggregate or in light weight cement. These markets are difficult to penetrate and are unlikely to expand very fast. The control of fly-ash is well developed technically, either by using electrostatic precipitators or filters. Electrostatic precipitators can remove 99.9 per cent of particulate matter and are normally installed in power-stations.

The control of gaseous effluents is most highly developed for removal of sulphur dioxide. The extent of removal is dependent on the physical and chemical characteristics of the sulphur in the coal. In regions where coal washing is extensively practised but not for the purpose of maximum sulphur reduction, the potential exists to upgrade this cleaning practice to remove sulphur. Lime/limestone wet scrubbing may now be considered a commercial technique though operation problems still occur (associated with clogging and scaling). These units do, however, produce larger volumes of difficult solid sludges. These units do, however, produce larger volumes of difficult solid sludges. For this reason, land use restrictions may prevent the application of lime/limestone systems in regions of high population density. The development of regenerable systems, in which a saleable sulphur byproduct is produced, is moderately well advanced. Two systems, Wellman-Lord and magnesia scrubbing, are considered ready for commercial application for coal-fired boilers but they are not yet proven. Other systems are only at the pilot plant stage.

No post-combustion treatment systems suitable for small industrial boilers have yet been developed, though the use of fluidized bed boilers will allow control during combustion using limestone injection. (See section on fluidized bed combustion).

Nitrogen oxide control may be partly achieved by optimising boiler conditions to lower the amount of atmospheric nitrogen fixed during combustion. Post-combustion control is derived from a number of scrubbing systems, but is presently unproven on coal-fired boilers. Fluidized bed combustion is also expected to give low emission levels.

The technical control of carbon dioxide (which is not now required but which is being examined for possible climatic effects) has not proceeded beyond paper studies.

The control of trace emissions is technically unresolved and may prove a very intransigent problem if very low emissions are required. Most work is currently directed to establishing whether trace elements are retained in the ash or emitted.

Cost/Time Considerations

Control costs are very site dependent but an elaborate scheme might cost up to $6/ton of ash. Fly-ash costs depend upon the degree of control required. An ESP system to remove 99 per cent of ash would cost about $15/kW for a large power station.

Sulphur control costs range from $60-100 kW for scrubbing systems, not including waste disposal, and may use 5 per cent of gross power output and add 4-5 mills/kWh to electricity costs. Throwaway lime/limestone systems are available now. Further improvements aimed at improving the reliability of the system and the quality of the waste, will occur over the next 10 years but no significant cost reductions are to be expected. Regenerable systems will become widely available within 5-8 years. They will not reduce system costs except where higher waste disposal costs are otherwise incurred.

The cost of nitrogen oxide controls is speculative, but probably control could be achieved together with sulphur oxide control at little greater cost. Systems will be available in 8-10 years for coal-fired boilers.

The cost of controlling carbon dioxide, if necessary, or trace elements is quite unknown but is likely to be very high.

LOW-BTU GAS FROM COAL

Introduction

Gasification of coal has a history spanning several hundred years having its early beginnings in the production of town gas for industrial and domestic use. The process involves partial combustion of coal in air and steam such that the products consist primarily of the volatile components of the coal (driven off by the combustion heat) and further reaction products of water vapour with hot carbon (hydrogen and carbon monoxide). When air is used as the gasifying medium the product gas will obviously contain a significant amount of nitrogen and for this reason the heating value is rather low, of the order of 1.2-1.7 million calories per cubic meter (130-190 BTU per cubic foot).

Objectives

Gasification of coal to produce a flammable product is attractive for a number of reasons. The gases can be easily cleaned to remove sulphur and other objectional materials thus making the subsequent combustion of the flammable fuel gas a cleaner process capable of meeting requisite environmental standards. The clean fuel gas produced makes design of combustors or boilers much simpler and straight forward and it can be easily transported. Yet another advantage gained by gasifying (rather than directly burning coal in a utility boiler) is the fact that some of the current designs for gasifiers permit the use of a wide range of coal quality and types, thus expanding the fuel markets considerably. It is important to point out that low-BTU gas is suitable for use as an energy source only near its point of production because its low heat content makes it uneconomical to transmit long distances; say, greater than 1 or 2 miles. The prime reason for considering low-BTU level gas today is that it can be utilized in a combined cycle power plant to generate electricity at higher efficiency than today's conventional coal fired systems or as an industrial energy source for process heat and steam raising. It can also be used as a reducing agent for process metallurgy and for industrial non-fuel applications such as ore reduction and synthesis gas for chemical feedstocks and methanol production.

State-of-the-Art

As mentioned above, gasification of coal is not a new development and there are a number of gasifiers presently available in large commercial sizes. The most well-known of these is the Lurgi gasifier presently being used industrially in very large gasification plants in South Africa. A 180 MWe system using Lurgi gasifiers has been demonstrated in West Germany and other low-BTU gasification installations are presently in operation throughout the world. In addition to these commercially proven gasifiers, there are a large number of new gasification concepts which are currently being developed in several countreis. These new developments promise moderate increases in efficiency, an ability to handle a wider ranger of coal type and perhaps eventually improved operating simplicity and reliability.

Some major problem areas which require significant development effort are pressurized coal feed, particulate removal from the gas product, removal of sulphur compounds at high temperature, and the design of high temperature heat recovery systems.

Cost/Time Considerations

Capital cost estimates for a complete 800 MWe integrated gasification/combined cycle power plant are $800/kWe, producing electricity at 37 mills/kWh with an overall coal-to-electricity efficiency of about 40 per cent. Costs of fuel gas produced by low-BTU gasification are included in Table G-2. The construction period for a large plant is estimated at about 5-7 years; smaller size units for industrial or small generating station use would require less.

HIGH-BTU GAS FROM COAL

Introduction

Gasification of coal can be carried out in oxygen rather than air and when this is done the fuel value of the gas is increased by eliminating nitrogen. This produces a gas of medium calorific value (2.6-3.6 million calories per cubic meter; or 290-400 BTU per cubic foot) which can be used to advantage as an alternative to low BTU gas within a distance of about 25 miles of the gasification plant itself. Such gas can be further upgraded to produce a gas with a heating value of approximately 9 million calories per cubic meter (1 000 BTU/cubic foot) called high BTU gas.

Purpose/Objectives

High BTU gas is similar to natural gas and therefore is a direct substitute for it. It is thus often called substitute natural gas (SNG) and can be transported and burned like natural gas. Medium BTU gas (mainly a mixture of carbon monoxide and hydrogen) is an excellent feedstock for the synthesis of a wide range of fuels and chemicals.

State-of-the-Art

As mentioned in the previous section on low-BTU gasification, the concept of gasification of coal is not a new one and thus some commercial development and exploitation of this concept has already taken place. The Lurgi and Koppers-Totzek gasifiers are the most widely used commercially today. Both these systems are presently in operation in South Africa on a large commercial scale, and the synthesis gases produced are used in the production of gasoline, ammonia, and heavy hydrocarbon liquids.

Thus, the development of medium and high BTU gasification processes has already reached a commercial stage in some instances. However, such plants in general do not produce a product at a cost competitive with alternative energy supplies so that an active development effort is underway in several countries with the objective of reducing the cost and improving plant reliability.

Probably the new process developed furthest towards commercial realization is the British Gas Corporation/Lurgi gasifier in Westfield, Scotland. The detailed advantages and problems associated with these new processes are similar to those for low-BTU processes.

Cost/Time Considerations

Recent capital cost estimates for a plant producing 250×10^6 cubic feet of SNG per day range from 1.1 to 1.8 billion dollars (1976 basis) producing SNG at \$2.7 to $4.7/10^6$ BTU. The construction period for a typical SNG plant is estimated to be about five to seven years.

PREPARATION OF LIQUID FUELS FROM COAL

Introduction

The weight ratio of carbon to hydrogen in coal varies from 12 for a typical lignite to 20 for typical bituminous coal. By contrast, petroleum-derived products range from 4.5 for propane to 10 for a heavy residual fuel oil. Thus the conversion of coal to a liquid involves either the addition of hydrogen or the removal of carbon. Processes using the former route include direct hydrogenation and solvent extraction-hydrogenation and also gasification-synthesis, which indirectly adds hydrogen. Pyrolysis processes are an example of carbon removal.

Objectives

The objective of coal liquefaction is the production of environmentally acceptable liquid fuels and feestocks which can be used for blending with or as a substitute for petroleum-based materials. Liquid fuels and feedstocks can be more readily handled than coal and have higher thermal values per unit volume. This is particularly significant for automotive purposes (both gasoline and diesel fuels, where significant premiums over values based on thermal content already exist). Other areas of particular interest are petro-chemical feedstocks, gas turbine fuels, and " clean" boiler fuels where, again, significant premiums over values based on thermal content can be expected.

State-of-the-Art

While large-scale coal liquefaction plants have been built in the past using the processes outlined above, only one plant, that of Sasol in South Africa, is currently in operation. This plant uses the gasification-synthesis route. A further plant is now under construction for Sasol. A large demonstration unit (545 tons/day) is being built for the H-Coal direct hydrogenation process (USA) and detailed designs are proceeding for similar large demonstration units for the Exxon donor solvent (solvent extraction-hydrogenation) and Cogas (pyrolis) processes.

The major objective of all process development is to reduce the cost of the fuel produced, compared with the existing performance of synthesis routes. However, there remain technical problems with all routes apart from gasification synthesis, in the separation of undissolved coal and residual mineral matter from coal liquids and solvent. The development of high capacity, high differential slurry pumps and other mechanical equipment, such as pressure let-down valves handling corrosive and abrasive mixtures of solid liquids and gases is also required.

Cost/Time Considerations

The construction period for a typical liquefaction plant is of the same order as that of a high-BTU gasification installation; that is, about 5-7 years. A typical coal liquifaction plant could have a thermal output of 28×10^9 BTU/hr at an average cost of $3 per 10^6 BTU and would require a total investment of approximately 1.7×10^9 (1977 basis).

FUEL CELLS UTILIZING COAL DERIVED FUELS

Introduction

The fuel cell is a device which withdraws the energy of a spontaneous chemical reaction at specially-designed electrodes and converts it to electrical energy. Its basis of operation is fundamentally the same as that of an ordinary battery, but the chemical reaction in this case is the combustion of hydrogen (or certain hydrocarbon fuels) using oxygen or air and the temperature of operation is usally quite high (60-190 °C for low-temperature fuel cells; 600-1 000 °C for high-temperature ones). Like ordinary batteries, it produces direct current (DC) electricity.

Fuel and oxidant are fed to the cell at rates proportional to the desired level of power production. Reaction takes place at specially designed gas-absorption electrodes which often catalyse the reactions.

Objectives

The feature of fuel cells which makes them particularly attractive for power production is that, not being Carnot-cycle devices, they can convert fuels to electrical energy at efficiencies approaching 90 per cent, while conventional steam cycles are limited by thermodynamic laws to efficiencies of about 40 per cent. In addition the cells are quiet, clean, and should be relatively maintenance free.

State-of-the-Art

Today, the largest operational fuel cells have a capacity of only a few MW and experience thus far indicates a number of weaknesses in the present state of development of this technology:

1. *Fuel purity:* the hydrogen/hydrocarbon fuel must be exceptionally pure (especially in low-temperature devices) in order to avoid poisoning the electrode surfaces and thereby significantly reducing reaction rates. Carbon monoxide is a major component of most coal conversion product gases and, unfortunately, is also such an electrode poison.
2. *Low Power Density:* the amount of power produced for a given size (surface area) electrode needs to be increased if economics are to be attractive.
3. *Cell Life:* experience so far indicates that electrode poisoning reactions, electrolite degradation, and similar problems prevent attainment of the long cell life (greater than, say, 10 years) necessary for economic power production.

As with other direct conversion methods, the electric power produced by fuel cells requires costly inversion of direct current (DC) to alternating current (AC) before it can be used by present distribution systems.

Cost/Time Considerations

This technology might be considered well-developed if one looks only at small, low-temperature devices; however, when considering major power production for utility distribution, one must conclude that the concept is not yet fully-developed. Recent estimates of power station capital costs based on fuel cells range from 425-900 $/kWe, but these estimates include assumptions concerning integrated gasification plants and waste heat utilization designs which are not yet proven. Estimated fuel cell life is optimistic as well.

The high-temperature systems, which could be coupled with conventional steam bottoming cycles, are the most efficient (near 52 per cent) but it is unlikely that any significant amount of power will be produced by this means before the year 2000. Efficiencies calculated for low temperature systems are not nearly as good as this, ranging between 12 and 32 per cent, but they do have the advantages of reduced temperature operation and the associated reduction in maintenance costs and operational complexity.

COAL-FIRED MAGNETOHYDRODYNAMICS (MHD)

Introduction

The concept of power generation by magnetohydrodynamics is based upon the fact that an electrical potential difference is created in a conducting fluid which passes through a magnetic field, inducing the flow of an electric current. The conducting fluid can be either a high-temperature gas (or plasma) or a liquid metal, but liquid metal systems are not generally considered to be near-term technology nor do they offer the performance potential of open-cycle gas systems. In either case, superconducting magnets are required.

Objectives

The most desirable feature of an open-cycle MHD generator is its ability to extract significant quantities of usable energy from high-temperature ionised gases (2 200-2 500 °C) without the use of a turbine. These gases, after passing through the MHD generator, exit through a diffuser at about 1 900 °C and can then provide the heat necessary for operation of conventional steam-driven turbo-generation systems. Resulting overall " coal pile to bus bar" efficiencies of the combined systems exceed 50 per cent. This advantage is not gained without penalty, however, and the complexity of MHD generators is such that construction times are expected to be quite long. Furthermore, only DC current can be produced by this means, requiring construction of costly AC/DC converter units.

State-of-the-Art

Development of MHD technology is confined to the laboratory at present, with most experimental studies being performed on single channel benchscale units. Among the most significant technical problems to be solved before commercialization can be considered feasible are:

1. *Gas/plasma conductivity enhancement and seed recovery :* injection of an easily-ionised alkali metal carbonate " seed" material (potassium or a cesium) to improve conductivity of the MHD working fluid makes necessary efficient recovery/regeneration techniques. Seed material must first be separated from ash and slag and then be regenerated, and recovery at better than 95 per cent efficiency is thought to be necessary for economic operations.
2. *High-temperature heat exchanger design :* the large amount of thermal energy in the MHD generator exhaust gas requires substantial recovery for reasonable thermodynamic efficiency, which means full utilization of combustion air pre-heaters at temperatures in excess of 1 400 °C — significantly beyond the limits of present technology (especially when " dirty" flue gases are considered).
3. *Materials of construction :* in addition to the materials problems implied in item (2) above, MHD combustors and generator duct materials must withstand attack by high-temperature gas components and slag. Additionally, the costs of superconducting magnet systems, AC/DC converter systems, and other sub-systems unique to this technology need to be reduced substantially.

Cost/Time Considerations

It has already been pointed out that MHD power generation concepts are at a very early stage of development and for this reason cost projections for such systems are necessarily rather speculative. Recent studies, however, suggest that large MHD systems (greater than 500 MWe) are necessary for good economics and will require an investment of between U.S. $ 650-1 100/kWe. MHD power station construction times will be 20-30 per cent longer than for conventional systems owing to their complexity.

As a result of the technical and economic factors listed above, one can assume that no significant amount of power will be produced by this technology before the year 2000.

TRANSPORT OF COAL

Introduction

At present, the energy content of coal is almost entirely transported in its original solid form or as electricity from mine-mouth generating stations. This section discusses development work in the area of coal transport only.

Purpose/Objectives

The objective of coal transport R & D is to reduce the cost and/ or improve the reliability of coal transport and at the same time reduce the environmental impact.

State-of-the-Art

Coal transport is either overland, across oceans, or on inland and coastal waterways. Overland transport is mainly by conveyor belt and road vehicles for short distances and by rail for long distances. In the last ten years there has been a considerably increase in the capacity of bulk carriers for ocean transport; the largest are now 120 000 deadweight tons. Inland and coastal waterways carry barges and small ships, which can be self unloading.

Table G-3   Coal Conversion and General Utilization Technology Costs : A summary

	Largest Unit in Commercial Service	Projected Overall Coal-to-Fuel/ Electricity Efficiency (%)	Estimated Time for Plant Construction (yrs)	Recent Estimates of Capital Costs[1] (1976 $) /kw	Recent Estimates for Cost of Electricity or Fuel[2] (mills/kwh unless otherwise stated) (1976 $)	Earliest Likely Date for Extensive Commercial Use[3]
Conversion Technology :						
Low BTU Gasification[4]						
Power generation/combined cycle	170 MWe	39-43	5-7	800	37	1985
High BTU Gas (SNG) production	4 × 10⁶ m³/day	59-63	5-7	($ 210/m³/day)	($ 3.5 million BTU)	1985
Liquefaction	6 000 b/d	63-78	5-7	($ 42/b/yr)[5]	($ 2.8 million)[5] BTU	1990
Fuel Cells (Coal Fired)	(none)	12-52	4-6	425-900	40-55	1990
Magnetohydrodynamics	(none)	45-52	7-9	650-1 100	—	After 2000
General Utilization Technology :						
Mining	—	—	10-12 (underground) 3-5 (surface)	—	—	—
Fluidized bed combustion[4]						
Atmospheric with steam cycle	30 MWe	35-37	5-7	620	32	1985
Pressurized with combined cycle	(none)	38-42	6-8	710	34	1990
Environmental protection[6]	—	—	—	15 (fly-ash) 60-100 (sulphur)	5-8	—

1.   Capital costs given are those calculated for central station, base-load power generation facilities or for large plants presently in operation. The values shown are order-of-magnitude only and wide variations (even for conventional systems) will be observed, depending upon the estimator, 1976 dollars are used. Gas and liquid production facilities are based upon average fuel value data and are quoted in terms of daily gas production rates or annual oil production rates.

2.   Cost of electricity/fuel values are based upon the quoted capital costs and construction times. Interest on capital during construction is taken as 6.5% and a return on investment over a 20 year plant life of 18% is assumed. Coal feed is assumed to cost $ 0.95/million BTU.

3.   This means general commercial availability with performance guarantees.

4.   The conclusion to be drawn from these figures is that while there is evidence that each of these newer technologies for electricity generation is cheaper than conventional generation with sulphur scrubbers, there is yet no clear evidence that one is significantly cheaper than another. In particular the cost difference shown between the two types of fluidised combustion technology reflects the fact that under the convention shown in Note 2, the extra efficiency of conversion to electricity obtained from pressurised fluidised bed combustion does not economically outweight the cost of the extra investment compared with atmospheric fluidised bed combustion. With a higher coal cost and a lower charge on capital investment (for example European circumstances), the cost difference shown on these figures would be expected to reverse.

5.   Estimate of the cost of a low suphur liquid suitable as an industrial power generation fuel. For higher value liquids after refining, for example, substitute gasoline for transport, significantly higher costs would apply, more than the SNG cost of $ 3.50 per million BTU's.

6.   Cost of flue gas desulphurisation including disposal of the waste.

The main new technology is coal slurry pipelines. One such pipeline transporting 5 million tons per year over 440 km has been in operation for some years in the United States. Very little R & D is now being done on coal transport.

Cost/Time Considerations

Overall transport costs represent up to 75 per cent of the total cost of coal delivered to conversion plants. A significant cost is always that of transferring from one mode of transport to another. Slurry pipelines have roughly the same overall costs as rail transport, but a much higher proportion of pipeline costs are capital charges. The costs of ocean transport have been decreasing with the economies of scale of large ships, but it appears that the limit of these economies is now being reached because of the cost of enlarging ports.

Railways are restricted by the topography of the route and there may be environmental objections to greatly increased traffic from residents along the route. Slurry pipelines need a ton of water for each ton of coal and so are restricted to regions with adequate water supply at the origin and there are considerable environmental problems in disposing of the dirty water at the destination. Slurry pipelines can also have legal problems in obtaining right of way.

It is not generally considered that transportation will be a restraint on the potential utilization of energy from coal though overall transport costs will prevent some mining districts from being exploited. Slurry pipelines, it appears, will be constructed in preference to railways when no railway system already exists. There is likely to be an increase in the transport of coal-derived energy in the form of gas, synthetic crude and methanol.

H

ENVIRONMENTAL, HEALTH
AND SAFETY CONSIDERATIONS*

Expansion of coal production and utilization could result in greater environmental stresses than from today's production levels. Evolving policies to foster expanded coal use will have to consider means to diminish the environmental, health and safety impacts of each phase of the coal cycle: mining, coal cleaning, coal transportation and storage, conversion to synthetic fuels and combustion. Experience has shown that the social and economic benefits of reduced adverse impacts frequently justify incurring the control costs. However, the increasing costs of control may in certain instances be such that a closer balance will need to be struck between environmental and energy objectives.

The environmental concerns arise out of perceived specific impacts on the environment of production and use of coal. These impacts can be divided into hazards to health and safety, deterioration of water quality, material damage and soiling, visibility reduction, weather modification, climatic changes, agricultural and plant ecosystem damage, and land usage and alteration. The effect on health and safety may be the most important ultimate impact to be considered in the development and use of coal. Mining poses the greatest threat to the labour force while combustion may present the greatest health and safety risk to the general public in the form of air pollution and by toxic trace elements released into the air, the water supply, or the food chain.

The other main concerns are land reclamation after mining, and the strain on water resources in some coal-producing regions.

More work is needed to identify more precisely the critical environmental risks from greater coal usage and to perfect control measures. Extensive research and development programmes are particularly needed to upgrade cost effective air pollution control. While some technological constraints prevent environmentally-sound development and use of specific coal resources, there is no technological constraint yet identified so great as to negate greater coal usage in general. Indicative costs of pollution control technologies for the whole coal cycle are presented in this chapter.

The possible effects of carbon dioxide (CO_2) emissions released from fossil fuel combustion on the global climate could perhaps be an important environmental problem facing expansion of coal combustion. The CO_2 problem could be a truly international one because of the migration of pollution. A careful study of this possible problem is needed.

Increasing public awareness of the environmental and health consequences of coal production and utilization has led most OECD countries to introduce pollution control legislation in one form or another. The increasing stringency of such legislation could act as an increasing constraint on the expansion of coal utilization, if only in delay in adopting standards and because of public interventions against coal projects. These constraints and delays can be minimized by the timely development of well integrated national energy and environment policies and administrative arrangements consistent with prompt public resolution of environmental concerns about specific projects.

---

* This chapter was written with the assistance of the staff of the Environmental Directorate of the OECD Secretariat.

COAL MINING

Surface Mining

Of all the present environmental problems confronting the coal industry, reclamation of strip mined lands is one of the most important and can be expected to assume increasing importance in those OECD countries with large strippable coal reserves.

The environmental effects of strip mining and the ease of reclamation vary with the particular mining region and the mining method used. Reclamation is most difficult for " contour mined" mountain areas and easiest for flat " area mined" land. Under some circumstances, e.g., in very arid regions or on steep slopes in mountainous regions, satisfactory reclamation may be difficult.

If mining is undertaken without regard to its environmental effects, the land is left scarred and unusable. Streams can be polluted by silting and acid mine drainage. Another concern is the disturbance of underground hydrological systems that can be caused by mining. The continuity of the aquifer (the underground water bearing strata) may be disrupted thereby affecting the ground water resources of communities. These problems of acid mine drainage and hydrological disturbances are not easy to overcome economically, but the technology, although difficult, is known.

It is possible with current technology to restore most mined lands to their original condition or even to an improved condition. However, in many cases, the desired end use of the land may not be identical to the original use.

The estimated costs for strip mine reclamation are presented in the Table H-1 below. Actual costs will depend on the thickness of the seam being strip mined and to a lesser extent on the thickness of the overburden.

Table H-1　　**Strip Mine Reclamation Costs**

Area	Reclamation Cost Per Acre (1977 $)	Reclamation Cost Per Ton (1977 $)
US Western coal (thick seam)	$ 3 000	$ 0.16
US Central coal	$ 5 000	$ 0.89
US Eastern coal (thin seam)	$ 8 000	$ 2.91
Western European average	$ 5 000	

Throughout the coal cycle water requirements may present a major resource problem in certain areas of the OECD. Water requirements for revegetation and for dust suppression during mining are not very large and have been estimated at 20 to 60 litres per ton of coal produced. However, in arid regions, such as U.S. Western districts, even that quantity of water could present a supply problem.

In countries where large quantities of coal are strip-mined — the United States, Australia, Canada and Germany — legislation exists for the control and reclamation of strip-mined areas. The major producer, the United States, has recently enacted national legislation (the Surface Mining Control and Reclamation Act 1977) and this type of comprehensive legislation may signal a future regulatory trend in other countries.

The U.S. legislation stipulates that today's coal mining must also bear the cost for reclamation of old abandoned mines. For this purpose, a reclamation fee is assessed in the amount of $0.35 per ton of hard coal and $0.10 per ton of lignite produced from surface mining.

Deep Mining

The environmental effects of primary concern for underground mining have to do with surface subsidence and acid mine drainage. There are two possible approaches to the surface subsidence problem. The first approach is not to allow subsidence at all as in South Africa. This requires that pillars of coal are left unworked for roof support and so a lower proportion of recoverable reserves is extracted. The second approach, which is environmentally less desirable, is the practice used in the United Kingdom to recognize a right to let down the surface and to introduce legislation which rationalizes claims for subsequent damage to surface property.

Typical costs of the damage from mine subsidence have been estimated in the range of $1.00 to $5.00 per ton of coal, with an average cost of $1.50 per ton. While present coal producers are often not required to bear the costs of past damages, future legislation may impose this cost on future production.

Presently in the United States, the room and pillar method recovers about 57% of the coal in place at an average production cost (excluding environmental and occupational health and safety cost increments) of $9.50-10.60 per ton production. However, if an additional 5% to 20% of the coal must be left in place to prevent subsidence, the incremental mining cost is between $1 and $2 per ton produced. Additional operational cost is incurred, of course, in the reduced recoverability of the coal in place.

The cost to the U.S. industry of the abatement of past damage from abandoned underground mines in the United States has been put in billions of dollars. The cost of reclamation of these mines will be partially borne by a reclamation fee on active mines of $0.15 per production ton.

Occupational health and safety in underground mining is constantly being improved. These actions are producing benefits both to society and to the energy economy by increasing the ability of the mining industry to attract new miners. Of course, these health and safety measures have increased the cost of coal production. Black lung protection and insurance measures in the United States have added an estimated $2.00 per ton to the cost of coal.

The safety requirements of the Mining Enforcement and Safety Administration (MESA) in the United States, while reducing productivity and increasing the labour cost of underground mining by $4.00 per ton, considerably improved occupational safety conditions in mining.

Water Pollutants from Coal Mining

Perhaps no major industrial water pollution problem is as complex or will be more costly to remedy than acid mine drainage. Pyritic materials in mines and storage piles for coal or refuse, exposed to moisture and oxygen, are oxidised to form sulphuric acid and dissolved iron sulphate. These discharges destroy aquatic life in streams, make water corrosive and unfit for industrial use, react with alkaline substance in the earth to harden the water, and deposit some undesirable substances along watercourses.

An important facet of any programme combatting water pollution is prevention of the problem at its source, and it involves the following practical control measures:

- drainage control to prevent water entering the mining area, and rapid removal of any water present;
- proper disposal of sulphur-bearing materials to prevent contact with water;
- sealing up abandoned mines to prevent water entering the sulphur-bearing soil; and
- chemical treatment of mine drainage under certain circumstances.

The U.S. Environmental Protection Agency (USEPA), in developing effluent discharge guidelines for mine drainage during 1977, has estimated the following maximum compliance costs (in 1975 dollars) for the use of best practicable technology currently available and for the improved technology which will be economically achievable in 1985:

Table H-2   Mine Drainage Treatment Costs
$/ton in the United States

	1977 Technology	1985 Technology
Eastern Deep Mine	0.03 - 0.28	0.17 - 0.51
Eastern Strip Mine	0.24	0.28 - 0.40
Central Deep Mine	0.02	0.06
Central Strip Mine	0.30	0.34
Western Strip Mine	0.10 - 0.12	0.12 - 0.15

Atmospheric Emissions from Coal Mining

Coal dust, especially respirable fine particles, is the primary air emission associated with coal mining. In order to suppress dust, occupational health and safety regulations often prescribe the sealing or spraying of major access roads, parking areas, and coal transfer points; the installation of dust collectors; and the reclamation of disturbed areas. The cost of these measures is estimated at $0.10-$0.20 per ton of coal produced.

COAL CLEANING

Present coal cleaning practice involves crushing and washing of the coal, with the application of froth flotation methods in some instances. The disposal of coal cleaning plant waste is a world-wide problem. In general, 25% to 50% of raw coal mined is disposed of as waste. The wastes should be disposed of in a manner which minimizes the leaching of trace materials and soluble salts or which permits the collection and treatment of runoff. Similarly, steps need to be taken to prevent dusting and spontaneous combustion. The abandoned waste piles from old preparation plants have been a major source in the United States of fine coal sediment and discharges and also of spontaneous combustion.

Recirculation and treatment of wash water are integral parts of the operation of moden coal cleaning plants. Closed water circuits reduce makeup water, eliminate discharge to streams and allow for improved recovery of fine coal. For example jigging requires 6 000 to 8 000 litres of water per ton of coal processed. If the water is recirculated then the makeup water requirement will only be 120-150 litres per ton of coal processed.

The USEPA effluent guidelines for new coal preparation plants are based on closure of the water circuit for preparation plants and on water treatment for effluents from coal and refuse storage. Compliance costs for this are estimated to be $0.07 per ton coal cleaned (1974 dollars).

Thermal drying of washed fine coals is the largest single source of air pollution in coal cleaning plants where they are employed. They are very common in the United States and where coals are prepared for long shipment. Uncontrolled emissions of particulate matter may be in the range of 15 to 25 pounds per ton of thermally dried coal. Thermal dryers also generate gaseous pollutants including sulphur dioxide, nitrogen oxides, carbon monoxide and hydrocarbons. Air pollution control devices for these emissions (nitrogen oxide excluded) are readily availabel. The USEPA has estimated the cost of thermal dryer particulate control for the levels of control. To meet an emission standard of 0.07 or 0.16 g per dry Nm^3 for particulates, the costs of $0.06-0.07 per ton of cleaned coal or $0.05 per ton respectively would be incurred.

COAL TRANSPORTATION AND STORAGE

Inland coal transportation is generally by railway or waterway for longer distances, and by conveyor belt or truck for shorter distances. In the future slurry pipelines may become a more important mode of transportation. The principal impacts have been fugitive coal dust and the possibility of acid water drainage both from transportation and open storage. These problems are similar to those experienced in the preparation plant storage areas and should be solvable even during transportation at costs of less than $0.05 per ton of coal shipped.

Slurry pipelines pose a different problem because they require approximately one ton water per ton of coal. There is a question of whether enough water resources will be available for that use in certain producing regions. In addition, at the end of the pipeline, about two thirds of the water is separated from the coal but is contaminated with suspended solids and occasionally chemicals used for fine coal floatation which will require treatment. If best available technology were to be used, the cost of slurry pipeline water treatment could be $0.15-0.25 per ton coal of coal transported. These costs could be reduced in two ways, but recovering much less water from the pipeline: burning the slurry with greater water content, or by using thermal drying to remove a larger percentage of the water.

In the past, the operation of a coal terminal has generally presented aesthetic problems with unsightly coal docks and coal fragments covering the bottom of surrounding waters. Depending upon recreational values and aesthetic needs in the vicinity of existing and potential coal terminals, more of the operation may need to be enclosed and made cleaner as world trade expands.

The storage of any coal can present problems of spontaneous combustion from reactions between the coal and atmospheric oxygen at ambient temperatures. High volatile, low rank coals that tend to weather are more likely to spontaneously combust and should not be stored if it can be avoided. The coal may be compacted in layers in order to exclude oxygen. For coals that are to be transported by barge or ship for any period, proper compaction to prevent rapid oxidation may be difficult to accomplish.

COAL UTILIZATION

The major environmental problems arising from coal combustion are creation of an ash waste and emission of air pollutants. The pollutants emitted (sulphur oxides, nitrogen oxides and particulates) cause health effects on the general public and impacts on crops, materials, aesthetics, etc. Because these pollutants can be transported over long distances they may contribute to lake acidification. Problems arising from emission of trace elements, and organic chemicals are discussed in the health section below.

Thermal discharge from all large power stations may also cause problems such as modification of the aquatic environment and destruction of aquatic life if discharged into water, or local climatic changes if discharged into the atmosphere. Water consumption and average costs for the various thermal discharge technologies for fossil-fired plants are shown in Table H-3.

Table H-3   **Costs of Several Thermal Discharge Options**

	Water Consumption	Capital Cost	Operating Cost
	kl/ton Coal Burned	$/ton/yr Capacity	$/ton of Coal Burned
Once-through Cooling	0 (evaporation	0.90	0.00
Cooling Ponds	5 excluded)	1.80	0.50
Evaporative Cooling Tower	7	2.10	0.80
Dry Cooling Tower	0	6.00	7.00

The decision whether to use cooling towers depends on the availability of water and total cost. Environmental concerns will usually indicate that cooling towers be used.

Collected ash (bottom ash and fly ash) presents a disposal problem. A small fraction may be converted to commercial use. The remainder is normally pumped to settling ponds where it settles out of the transport water. Although disposal costs are very site dependent, typical costs can be exemplified by a 500 MW power plant requiring 75 acres of settling ponds over its 30 year life. The investment cost of the disposal system and lined pond (to prevent run-off or leaching, of for instance heavy metals and acidity, into ground water) would average $9/kW (land at $4 000/acre represents less than 10% of the investment). Annual cost would amount to $1.80/kW or $0.70 per ton of coal burned.

Electric power plants using coal will call for more land as compared to oil-fired plants. Furthermore, land requirements will be larger (maybe 10 times greater) where ash or sludge ponds are needed. Larger land needs by utilities will necessitate better siting procedures, assessment of socio-economic impacts and involvement of the public in the decision process.

EMISSION CONTROL TECHNOLOGIES

Particulates

To reduce particulate emissions to acceptable levels coal-fired power plants need to install high efficiency collection equipment such as electrostatic precipitators or venturi scrubbers. Modern designs of electrostatic precipitators on 3%-5% sulphur coal are capable of removal efficiencies greater than 99% with an investment cost of about $15/kW and annual costs (including capital charges) of $2.65/kW, or $1.05/ton of coal burned for medium to large power stations. However, the small quantity of particulates not collected is a fine type which is respirable and perhaps harmful to human health. Electrostatic precipitators are most effective on high sulphur coal particulate and lose considerable efficiency if sulphur content of the coal decreases to 1% sulphur. Because capital and operating cost of precipitators can be 2 to 5 times higher and special conditioning systems may be needed for low sulphur coal, venturi scrubbers are often preferred. These scrubbers are capable of efficiencies above 95% at investment costs of $19/kW and annual costs of $5.65/kW, or $2.20/ton of coal burned.

Sulphur Oxides (Flue Gas Desulphurization)

Of the many flue gas desulphurization systems that now exist, lime/limestone wet scrubbing is now considered a commercial process although operational problems may still occur during shakedown and early operation. As a result of the large volumes of sludge produced by the process, land use restrictions may prevent the application of lime/limestone systems in regions of high population density. The capital costs of lime/limestone scrubbing has been estimated at $50 to 70/kW for 1.5% sulphur coal-fired power plants and $60 to 90/kW for 3.5% sulphur coal (1977 dollars). These cost ranges include suitable measures for waste disposal or the production of gypsum instead of sludge waste. Annual costs of these systems will range from $19 to 32/kW or $7 to 12 per ton of coal burned.

The development of regenerable systems, in which a saleable sulphur byproduct is produced, is moderately well advanced. Two systems, Wellman-Lord and magnesia scrubbing, are considered ready for commercial application for coal-fired boilers but they are not yet proven in large-scale power plants. These systems might become widely available within 5 to 8 years. They will not reduce system costs but may be preferred to avoid producing waste or where high waste disposal charges are incurred.

Nitrogen Oxides

For limiting the nitrogen oxide (NO_x) emissions from power plants, the control techniques that could be employed include staged combustion (30 to 70 per cent efficiency), low excess air firing (10 to 30 per cent efficiency), and flue gas recirculation (20 to 60 per cent efficiency). Efficiency varies from case to case and is dependent upon original combustion chamber design and method of operation.

The capital costs of nitrogen oxide control on new coal-fired boilers is about $3.00/kW, the operational cost is about $0.50/kW/year, or $0.20 ton of coal burned.

Fluidized bed combustion at atmospheric pressure, because of its lower combustion temperature, will reduce NO_x, formation levels comparable to the above systems. Pressurized fluidized beds may reduce NO_x emissions even further.

Fue gas cleaning systems for NO_x emitted from coal-fired combustion are under development in Japan. These systems are capable of removal efficiencies of nearly 90 per cent. Certain flue gas desulfurization systems are being developed to remove SO_2 as well. If a high degree of NO_x removal is required, which may be very likely in the 1980s, these combined systems will be the most practical and economic method of control. However, they will cost at least as much as regenerable SO_2 scrubbers and perhaps 50% more. Capital cost of $10-25/ton coal burned is a very rough estimate for a combined SO_2/NO_x gas cleaning system.

Emission Control Regulations

Air quality regulations in most OECD countries limit SO_2 and particulate emissions from coal combustion. Additionally, in a number of countries, limitations are placed on nitrogen oxides and hydrocarbon emissions. These pollutants are under study in most OECD countries and it is anticipated that more widespread regulations will be in force in the 1980s. With increased use of coal, emissions of trace elements could become a problem and some limitations on these emissions may be introduced after 1985. Such problems and potential regulations should be anticipated and cost-effective control technologies perfected so as not to restrict expanded coal utilization in the future.

Air quality regulations can be of two general types, those that specify certain ambient air quality goals and those that limit emissions from individual sources. Achievement of ambient air quality goals may be through emission limitations or through the dispersal of pollutants using tall stacks. However, where long range transport of pollutants or transfrontier pollution is a problem the use of tall stacks may not be a viable control option.

In the United States, national ambient air quality standards are the basis for setting limitations on emissions to the atmosphere. In addition, New Source Performance Standards (NSPS) are applied to new stationary sources for specific categories and plant modifications resulting in new emission levels. These NSPS are to be achieved through the application of the best available control technology which has been adequately demonstrated in plant operation.

The approach adopted in Japan is to limit emissions according to local conditions. A more stringent standard is applied to new or modified facilities and those that are located in congested areas where air pollution is serious. In regions where this type of standard has been difficult to apply, total emission control is employed.

In the Federal Republic of Germany "Technical Instructions for Air Pollution Control" (TA-Luft) issued on the basis of the Federal Emission Control Law contains source performance standards and ambient air quality standards. As to coal-fired power stations, SO_2, CO, and particulate emissions are subject to limits. These emission standards require installation of high efficiency collection equipment for particulates, burning of low sulphur coal, and flue gas desulphurization for large power stations.

In some other European countries regulations governing the release of SO_2 and particulates to the atmosphere are of the general type specifying stack heights for the dispersal of pollutants from large facilities. In addition, fuel sulphur content may be limited for areas designated as specially protected zones where the level of atmospheric pollution is high. Of all European countries, only Germany has stringent ambient air quality standards for nitrogen oxides.

SYNTHETIC FUELS

In determining whether synthetic fuels are an environmentally acceptable alternative to coal or oil, it will be necessary to examine all the environmental aspects of the combined conversion and combustion system.

Basically, the conversion plant will have to deal with the removal and disposal of ash, particulate, sulphur compounds, and nitrogen compounds, sometimes similar and sometimes different in magnitude and type from the power plant problems. About 15%-20% of the coal will be burnt for power generation producing pollution similar to power plants. The rest of the pollution will result from purification of the synthetic fuels as potential air pollutants, water pollutants or waste depending upon technology used.

Many uncertainties exist with coal-based synthetic fuel production and its environmental impact. For one thing, synthetic fuel processes are complex (up to 80 different classes of compounds of potentially hazardous substances may be present in coal conversion plants). For another, there is very little environmental data or experience available on pollutant releases from the different processes.

A major resource constraint will be the large quantities of water needed for all coal conversion processes for steam production, sulphur removal, etc. Water requirements for fuel quantities comparable to the fuel requirements of a 1 000 MW electric power plant is 30 billion to 200 billion litres per day.

Considering the conversion and combustion of the four general types of coal-based synthetic fuels, only one — high BTU gas — will have definite environmental advantages over coal combustion. High BTU gas has all the environmental advantages of natural gas. The other three — low BTU gas, solvent refined coal and coal liquefaction products — appear to have adverse environmental trade-offs as fuel alternatives to coal.

For the low BTU gas the particulate matter, the hydrogen sulphide, the ammonia, the cyanide and the other sulphur components can only be removed by conventional condensation and scrubbing techniques which will probably not be economically acceptable. At present, removal systems to operate at high temperature are under development but low BTU gas cannot be both environmentally and economically acceptable without these systems. Since the gas is composed mainly of CO, the gas itself is toxic. Leaks can be extremely dangerous and therefore sophisticated warning systems should be developed.

Similarly, with solvent refined coal and coal liquefaction products, adverse environmental tradeoffs exist although the sulphur and ash content of the products is generally much less than that of the original coal. For example, higher NO_x emissions than from coal combustion may result. Also, solvent refined coals and liquefaction products are suspected of being carcinogenic and toxic as a result of compounds within the material and further research in this area is required.

HEALTH AND SAFETY ASPECTS OF THE COAL CYCLE

Health, generally

Air pollution has long been recognized as a cause of illness and death as well as discomfort. The principal health effects are believed to result from inhalation of sulphur pollutants and particulates. The causal relation of severe air pollution episodes to increased mortality and morbidity is firmly established. However, the identity of the specific pollutants which are responsible is not clear, nor the degree to which the effects are reversible as pollution decreases.

Experts disagree as to whether an effects threshold really exists — one argument being that there is no conclusive or convincing evidence that repeated low level exposures will not lead to eventual respiratory impairment. It may be that simple dose response relations cannot represent the effects.

Health effects are very dependent on the geographical pattern of expanded coal utilization and its relation to the habitat pattern. They are also sensitive to daily fluctuations in the air pollution levels, wind patterns and other meteorological conditions. The estimates thus far made of the health hazards of emissions from coal-fired plants differ so greatly (varying from 0 to 100 deaths annually per 1 000 MW plant) that as yet they offer little public policy guidance on coal development.

New Health Risks

As sulphur emissions are decreased the relative importance of other pollutants (for instance NO_x and heavy metals), which are presently poorly understood, would increase; and at very low sulphur levels these other pollutants may cause the dominant health effect.

Emissions of polycyclic hydrocarbons (PAH) and other carcinogenic substances are causing concern because they are being emitted and the effects are also not well understood. The health impact from these pollutants originating in energy production is considerably less than the impact of cigarette smoking and auto exhaust.

Occupational Hazards

All energy cycles have impacts on occupational health (diseases, primarily black lung for coal) and safety (injuries and fatalities).

The most predominantly hazardous element in the coal fuel cycle is mining, even if intensive technical and legislative developments over recent years have begun to reduce the risks involved. Of course surface and deep mining have different levels of risk. To produce a given amount of coal in U.S. deep mining causes three times as many injuries and four times as many deaths as surface mining.

Hazard statistics from the EEC 1970-1974 for deep mining show high values. More recent statistics from the Federal Republic of Germany 1975 show per 1 000 MW (per annum) :

Injuries
Rehabilitation < 20 days	394
Rehabilitation > 20 days	292
Deaths	1.6

Occupational illness
Certain disablement	42
Total disablement	1.6
Deaths	2

At a certain cost modern technology seems to be able to bring these statistics down further. Estimations exist for 1990-2000 of 0.5 deaths in deep mining and 0.1 in strip mining per Mt coal (twice as many per 1 000 MW).

Major unresolved concerns include the understanding of health impacts, the impact of coal conversion processes, and the emission of trace elements such as mercury, lead, cadmium and uranium. As coal use is greatly expanded, additional research into these problems will be necessary to determine how best to protect the environment as well as occupational and public health and safety.

A SUMMARY OF ENVIRONMENTAL CONTROL COSTS

Table H-4 summarizes the indicative ranges of environmental control costs discussed earlier in this chapter. These cost estimates were developed by the OECD Environment Directorate from a variety of sources and groups of experts. The indicative estimates for stages of the coal cycle shown in the table are non-additive. It is impossible for the estimates to fully reflect the wide variety of site-specific factors, environmental regulations, and coal types that will influence actual costs. Equally important, it is unclear to what extent the range of costs presented in the table reflect the differences among (*i*) the cost of compliance implied by various degrees of regulation, (*ii*) actual cost differences among existing technologies, (*iii*) differences in the costing techniques employed by the experts and (*iv*) uncertainties about the eventual costs of technologies not yet fully perfected[1].

---

1.   The coal utilization control costs presented in this Chapter are not directly comparable with those presented in Chapter F without some further adjustments.

Table H-4 Indicative Cost Estimates for Specific Environmental Measures
$/ton of coal, 1977 U.S. $

COAL MINING/COAL CLEANING

	Contour Surface Mining (thin Seams)	Area Surface Mining	All Surface Mining	Underground Mines	Comments
1. Reclamation of Active Mines (Including Prevention of Mine Subsidence)	2.80 - 3.00	0.15 - 0.90		1.00 - 5.00	Higher for surface mining in steep sloped areas
2. Fee for Reclamations of Abandoned Mines			0.10 (Lignite) 0.35 (Coal)	0.15	U.S. legislation
3. Dust Control			0.10 - 0.20		
4. Mine Drainage Control	0.35 - 0.50	0.15 - 0.40		0.07 - 0.60	1985 technology
5. Occupational Health and Safety Requirements				6.00	
6. Coal Cleaning - Prevention of Runoff from Storage and Wastes			0.09	0.09	Per ton cleaned

COAL TRANSPORTATION

	By Rail	Slurry Pipeline	Harbours	Comments
1. Dust control, Prevention of Spills, Control of Runoff	0.05		Unknown	
2. Treatment of Slurry Water		0.15 - 0.25		Reduced by evaporating

COAL UTILIZATION

		Comments
1. Control of Waste Heat Emissions by use of Cooling Towers	0.80 (wet)* 7.00 (dry)**	*7kl/ton coal H_2O consumption **No H_2O consumption
2. Particulate Control	1.05 (electrostatic precipitator) 2.20 (Venturi scrubber)	
3. Control of SO_x	7.00 - 12.00 (lime/limestone FGD system)	Depending on coal type and specific regulation; including waste disposal.
4. Ash Disposal	0.70 (in lined ponds)	
5. Control of NO_x by Combustion Techniques	0.20 - 0.30	

I

COAL PROSPECTS IN NORTH AMERICA

THE UNITED STATES

Nearly 65 per cent of the recoverable coal reserves in the OECD region are in the United States, where coal is the most abundant fossil fuel available. The U.S. reserves are sufficient to meet significantly higher domestic and export needs for well beyond a century. But the extraction, transport, burning and processing of this coal raises large problems concerning economic, environmental and other social costs. Thus, greatly expanded use of coal will require a change of public attitude towards coal, a fuel which for the better part of this century was steadily displaced by oil and gas to gain greater convenience, lower cost and cleaner environment. While recent higher prices for oil and gas, and increases in coal productivity through surface mining, have again made coal cost-attractive for power generation and process heat, large unresolved policy issues remain.

MARKET PROSPECTS IN GENERAL

To meet the projected domestic and export demand for both metallurgical and steam coal foreseen in the reference case, low nuclear, the U.S. would need to increase the output of coal from 555 Mtce in 1976 to 837 Mtce in 1985, 1 013 Mtce in 1990 and 1 181 Mtce by 2000, for an average annual increase of more than 3 per cent for the last 24 years of this century. While this projection falls short of the National Energy Plan (NEP) goal of 1 051 Mtce for 1985, it is comparable to the Bureau of Mines' 1985 and 2000 estimates. If the combined domestic and export demand were greater it is estimated that U.S. production could reach as much as 1 300 Mtce by the year 2000.

The projections for the U.S. coal market in the reference case are based on the continuation of energy policies in place, including the mandatory coal conversion programme of the NEP, which has been accepted by Congress. Oil and gas price controls are assumed to continue but be phased out by 1990, at which time domestic oil and gas prices are assumed to have reached international price levels.

The electricity sector demands for coal, adjusted to slower economic growth rates projected by the Secretariat were partially drawn from the Department of Energy (DOE) projections for 1985 and 1990; these rise rapidly between 1975 and 1990. The present study assumes that nuclear power capacity will be only 100 GW and 140 GW in 1985 and 1990 respectively, due to periodic interventions and construction delay, compared with 130 GW and 201 GW projected by the DOE. While this appears to give scope for greater coal consumption in meeting projected electricity requirements, additional tonnage of coal is not included in the reference case and oil was substituted in its place, on the supposition that utilities would resort to oil-fired facilities normally used to meet peak load demands in order to buy time while additions to nuclear capacity are completed.

Utility coal consumption increases sharply after 1990 in the reference case. This results from a combination of slower electricity growth and slower than previously-expected nuclear expansion (both for low and high cases) projected for the period from 1990 to 2000.

Use of coal to make gas is restricted to relatively low levels (6 Mtce in 1985 rising to 36 Mtce in 2000), since costs of high BTU coal gasification are expected to exceed the energy price levels assumed in this study. The enlarged coal case, however, is likely to indicate a commercial application of lower BTU coal gasification, a consequence of accelerated government-supported technological development and partially subsidized commercialization of conversion processes.

Industrial coal demands are shown to grow, as a result of the economic attractiveness of coal (and the unavailability of gas) under conditions of continuing price controls on oil. Coal demand is expected to be partly reinforced by measures compelling conversion of large-scale industrial boilers to coal.

The metallurgical coal estimates were guided by projections by the Department of Energy up to 1990; no DOE projections beyond 1990 were available and those of an independent consultant specializing in the steel industry outlook were used.

Table I-1   **Production, Consumption and Net Exports of Coal in the United States**
1976-2000 (Mtce)

	1976	1985	1990	2000
Production :	555.4	837.3	1 012.6	1 181.0
Consumption :				
Metallurgical coal	76.8	87.9	91.6	91.6
Utility coal	369.6[1]	615.6	745.7	800.0
Other coal	53.8	65.8	96.0	160.8
Total	500.2	769.3	933.3	1 052.4
Net Exports				
Metallurgical coal	43.2	55.4	60.7	70.0
Thermal coal	10.3	12.6	18.6	58.6
Total	53.5	68.0	79.3	128.6

1. Includes 15 Mtce for electricity generation by the industrial sector.

SUPPLY ISSUES

In 1976, production exceeds the sum of consumption and net exports by net addition to stocks.

Coal Reserve Data

The U.S. Geological Survey puts " demonstrated " coal reserve base (proved plus indicated reserves or all the coal thought to be in place), at 395 billion metric tons[1], from 1975 data. But no more than about half the total of demonstrated coal reserves

1. U.S. sources usually report coal tonnage in short tons. These data have been converted to metric tons, but since the sources usually do not report the average BTU content of the coal tonnage, these tonnages cannot be reliably translated into tons of coal equivalents as used in OECD statistics (metric tons are *higher* than tons of coal equivalent, most likely in the range of 5-10%.

are considered " economically recoverable"; the convention has been to apply a rate of recovery of 50 per cent to underground mining and 80 per cent to surface mining, which yields a " recoverable" reserve base of about 235 billion tons.

Table I-2   **Demonstrated U.S. Coal Reserve Base by Sulphur Content and Potential Method of Mining**

Billion metric tons

	Sulfur range				Total
	<1%	1-3%	>3%	Unknown	
Underground :					
East of the Mississippi River	24	44	60	24	152
West of the Mississippi River	90	10	7	12	119
Total underground	114	54	67	36	271
Surface :					
East of the Mississippi River	5	6	13	5	29
West of the Mississippi River	62	24	4	5	95
Total surface	67	30	17	10	124
Grand total	181	84	84	46	395

Source: US Dept. of the Interior.

As can be seen from the above table, about 70% was coal amenable to higher-cost underground mining and 30% to surface mining; about two-thirds of the coal with 1% or lower sulphur content would require underground mining. Some 46 per cent of the coal is thought to be East of the Mississipi; but 84 per cent of lower sulphur coal is found in the West.

Quality characteristics of coal also vary markedly by region. Bituminous coal predominates in the East where the average calorific value is about 12 000 BTU/pd. whereas in the West lower-valued subbituminous coal (72 per cent of the West's total coal) and even lower-valued lignite (11 per cent) bring the average calorific value to 8 800 BTU/pd. Obviously, the greater the reliance upon the Western reserves instead of Eastern to satisfy a given energy requirement, the more tonnage must be mined and transported.

The adequacy of the U.S. official coal reserve data has been recently called into question, particularly those relating to economically recoverable reserves. The data problem has several facets: the data collection; the criterion for measuring recoverable coal; and the extent of development and therefore knowledge in various regions. The U.S. Geological Survey publishes estimates of physical coal endowment or " demonstrated" reserve base, while the US Bureau of Mines reports " recoverable" reserve base consisting of well-measured coal beds of specified seam depth and thickness, and by coal rank and region.

Unresolved questions about reserve data, however, do not impair the conclusion that U.S. coal reserves are more than adequate to meet foreseeable production needs for steam coal in this century and well into the next. Taking the Bureau of Mines' most conservative criterion to Western strippable coal, the projected U.S. production for domestic and export sales in the reference case of 837 Mtce in 1985 and 1 181 Mtce in 2000 would be sustainable from the total economic reserves, with considerable margin for expanded output for the additional export demand or significant manufacture of synthetic natural gas embodied in the enlarged coal case. A steady expansion, for instance, from the estimated steam and metallurgical coal output in 1976 to 2000 would result in an accumulative production of some 22 billion tce over the 24-year period, a small fraction of the estimated recoverable 235 billion tons of both Eastern and Western coal.

Leasing Policies

Leasing is a policy issue primarily to the extent that demonstrated coal deposits lie on Federal or State-owned or managed land. The Federal Government owns about 43 per cent of the surface rights and about 60 per cent of the recoverable coal reserves of the Rocky Mountains and Great Plains region. Eastern coal deposits, by contrast, are almost exclusively in private ownership. The Congress' Office of Technology Assessment estimates that 80 per cent of future coal development in the West will occur on Federally-owned land or deposits. While only about 40 million of the 555 million tons of coal produced in 1976 is from Federal leases, the Administration estimates production under existing leases will increase to 54 million tons in 1980 and over 136 million in 1985.

There are presently 15 billion tons of coal deposits under Federal leases and an additional 8 billion tons under Federal preference-right " leases" (leases awarded to those finding minable deposits on public land previously classified as having known deposits). But much of this tonnage is in scattered deposits with insufficient reserves, or it cannot be recovered economically with present technology, thus reducing the tonnage likely to be mined.

Acreage offered for lease from public lands was greatly reduced after 1968 and terminated by a moratorium in 1971, including prospecting permits for preference-rights leases. The moratorium, however, does permit leasing of tracts adjacent to leased land where these tracts are needed for optimal development of the acreage. In 1972 the Energy Minerals Activity Recommendation System (EMARS) was created to develop land-use plans to ensure consideration of environmental and conservation issues. The EMARS is being re-evaluated, especially with respect to non-producing leases (the Mineral Leasing Act of 1976 prohibits leasing public lands to anyone holding a non-productive lease more than 10 years). At the writing of this study a Federal review group was reviewing, on a regional basis, long-term coal leasing issues.

It is not clear that the moratorium on new leasing of Federal lands for coal development or other Federal land management has significantly impeded coal development up to now. An audit of half the outstanding leases, indicated that after a minimum of 7 years under lease, only 52 per cent were under some stage of development as evidenced by the filing of mining plans. While some of these holdings were scattered and not yet commercial due to an insufficient reserve requirement, the results of the audit suggest that lower demand expectations rather than restrictive leasing policy has been the main constraint. It is not even clear, given the preference expressed in the National Energy Plan for use of Eastern deep-mined coal by Eastern industry and reinforced by the effect of certain environmental requirements (discussed below), that any new Federal leases will be required to meet 1990 needs; much will depend upon the expansion of Eastern deep mines. In meeting needs for after 1990, however, continuance of the present moratorium on new leasing of Federal land for more than a few years could serve as a real constraint.

But if the moratorium on leasing were not lifted shortly, the longer lead times required to open Western mines than Eastern (estimated by the Bureau of Mines to be longer by 3-12 years for surface mines and 1-8 years for underground, mostly as a result of settling environmental requirements and disputes), it could deny an opportunity for new mines with large capacity to come into operation. Only with considerable reduction in lead times could the prospect of a shortfall in Western productive capacity by 1990 be avoided, or sooner in a setting of accelerating demand for coal exports.

Tilting Toward Eastern Mines

In 1975 the Western mines produced nearly 16 per cent of the nation's output. By 1985, the industry expected the Western share to rise to one-third, and this would be realized, in the estimation of the Department of Energy, whether the NEP was accepted or rejected by Congress. But the Plan states that the proposed " installation of the best available control technology" (BACT) in all new coal-fired plants, including those that burn low sulphur coal, would have the effect of " even greater use of high sulphur Mid-Western and Eastern coals". This requirement that all powerplants by 1983 install scrubbers regardless of the sulphur content of the coal burned was viewed by DOE as favouring Eastern deep mines (which, being closer to Eastern plants, could be delivered at a lower cost than Western coal).

The DOE estimates in Table I-3 show total demand for coal will be 180 million tons higher with the NEP than without, of which 100 million tons is expected to be mined in deep Eastern mines. Of the coal consumed by utilities, the sector affected by the emission control technology, the portion of high sulphur to total coal rated by sulphur content will rise from 51 per cent to 83 per cent. This is contrary to the previously-held expectations of industry, but consistent with the likely consequences of BACT in diminishing the competitiveness of Western coal by reducing the sulphur premium without changing the transportation penalty to distant Eastern utilities.

Table I-3   **U.S. Production of Coal by Region and Mining Process, in 1985 With and Without NEP, as Estimated by U.S. DOE**

Million metric tons

	1975	1985	
		Without NEP	With NEP
East			
Surface		256.6	265.4
Deep		383.2	487.8
Sub Total	488.0	639.8	753.2
West			
Surface		312.2	338.0
Deep		14.6	55.9
Sub Total	91.6	326.8	393.9
Total	579.6	966.6	1 147.1

Source: US Department of Energy.

In a study contracted for DOE it was observed that the effect of BACT could be that new powerplants built after 1983 will select coal on the basis of transportation costs rather than sulphur content; and this in turn will deplete the mostly Eastern high-sulphur coal reserves. ICF stated this would increase the price of such coal " slightly " because of the flatness of the supply curves of Eastern mines and the gradual phase-in of powerplants, the construction of which has not yet commenced. However, Eastern deep mines have contributed little, compared with Western surface mines, to improved productivity, because of higher labour intensity and more restrictive safety, health and work practices. A study done for the Electric Power Research Institute put the marginal costs of Northern Appalachian mines, mostly deep-mines, at 200 per cent higher than those of Western Northern Plains. Clearly, if Eastern coal costs rise considerably this will permit cheaper Western coal to overcome the cost disadvantage of BACT; in that event, the previously-expected balance between Eastern and Western production might be partly restored but at higher cost and some delay.

Table I-4   **U.S. Coal Consumption by Type, by Utilities in 1985 With and Without NEP, as Estimated by U.S. DOE**

Million metric tons

	Without NEP	With NEP
Low sulphur	91.6	135.3
High sulphur	434.6	354.7
Sub-bituminous	131.5	147.4
Lignite	59.2	59.0
Total	716.9	696.4

Source : US Department of Energy.

Mine Capacity Expansion

Several studies have been made to estimate planned expansion in coal mine capacity to 1985. The studies have suffered from a lack of good coverage on mines producing 180 000 tons or less per year, which account for nearly one-fourth of all U.S. coal production. Also, announcements of new capacity to be installed are often speculative and subject to change at any time; delays frequently occur because sales contracts do not materialize, or financing or environmental problems develop. Most such surveys estimate that total productive capacity in 1985 will range from 800 million tons to 1 billion tons, including replacement of retired capacity.

Coal Transportation Facilities

This section notes briefly issues concerning the adequacy of U.S. inland transportation of coal; issues of ocean transportation of U.S. coal exports are dealt with in Chapter E.

The NEP's proposed national energy transportation system is to " develop means to encourage use of energy supplies nearest to consuming markets ... and in order to reduce the need for long-distance transportation". Such a system if implemented would reinforce the effect of BACT in expanding demand for Eastern coal. Prior to the publication of the Plan the Department of Transportation surveyed the forecasts of major coal-carrying railroads and found that they not only anticipated greatly expanded coal rail traffic — by 43 per cent — by 1980, but expected 60 per cent of the growth in traffic to originate in the West, not the East. Part of the disporportionate share of the West in these forecasts reflects the greater tonnage of Western coal that must be moved because of its lower average BTU content.

Federal Railroad Administration, railroad officials and rail car manufacturing representatives believe that the industry has the capability of providing sufficient rolling stock to transport a large portion of the new coal required for levels of output comparable to projections of this study. Beyond 1985, more than enough time exists to acquire rolling-stock, improve road-beds and lay new lines; the signing of firm long-term contracts by mining companies provides sufficient notice and lead-time to the railroads.

The likely demand for services of the Western railroads will be partly affected by the role that slurry pipelines can economically play or will be permitted to play. The

Office of Technology Assessment (OTA) has published a controversial study of the economic viability of pipeline versus railroad (unit trains) transport of coal. Four hypothetical but quite plausible routes for slurry pipelines were chosen. Slurry pipelines were found cheaper on two of the lines, where distance was greater or terrain less difficult; railroads were shown to be cheaper where volume was smaller, users were dispersed, distances shorter, or terrain difficult.

The route most comparable to the exporting of Western coal was that from Gillette, Wyoming to Texas (closely resembling the route of the most ambitious slurry line under active promotion, from Wyoming to Arkansas). OTA found that the rail cost exceeded the pipeline cost by nearly 50 per cent or $ 2.80 a ton ($ 8.70 compared to $ 5.90, in 1975 dollars with a 12.5 per cent rate of return). The study concluded that the cost advantage of slurry pipelines are highly " route-specific". In any case, slurry pipelines cannot be laid until Congress grants to pipeline companies the power of eminent domain to cross railroad right-of-way, or alternatively, the companies obtain such power from each state crossed by the pipeline. Proponents must also obtain rights to undistributed water in the generally arid West; if this water is not available at the point of origin and must be piped from several hundred miles away, some slurry pipelines would not be economical.

Realization of the coal export potential of the United States, as well as the domestic coal conversion programme, will require not only an improved and expanded U.S. transportation network but a competitive one in terms of transport costs. Today these costs range from 25 per cent of the delivered cost of Eastern coal to as much as 75 per cent in the case of Western coal. Since barge traffic in the Mississipi north of St. Louis is congested and appears to offer little opportunity for expanded movement of coal, any cost-saving innovations in land transport mode, such as slurry-pipelines on select routes, could help hold down transportation costs and improve the competitive position of U.S. coal in foreign markets.

Community Resistance to Mining for Export

Concern has been expressed by European utility managers, as well as by a few American coal and utility officials, that some American communities will resist expansion of coal mining operations, especially surface-mining in the West, when done under contract to export markets, or even more so, if done by partially foreign-owned companies. These assertions usually rest on intuition or casually-formed impressions drawn from specific communities. This is an area of public attitudes that has not been well surveyed. The experience of communities with either export-oriented or foreign-owned companies have often shown that any early resistance ebbs with the creation of new jobs and tax revenues by these companies. Often the most vocal opposition comes not from the producing community, but potential domestic consumers who fear an advance in price or temporary shortage of the commodity because exporters are out-bidding domestic buyers (this concern rose, for instance, in the exporting of Northwestern softwood lumber). But the supply curve of coal — especially Western surface-mined coal — is fairly flat and export demand is not likely to bring about higher prices or shortages. Furthermore, utilities in the United States appear to have a greater differential between U.S. coal and other fossil fuels or even nuclear than foreign utilities and are thus in an advantageous position to bid against exporters for long-term coal contracts.

The nature of the transition to production for export could well improve public acceptance. Local communities would probably be more receptive to surface mining, especially in the West because of the environmental disturbance, if initial operations were begun by American-owned companies under contract to domestic, or better yet, local utilities or industrial users; once the community had accepted mining and was

satisfied with protection or restoration of the environment, expansion for export, even
by partially foreign-owned mining companies, would probably be easier. On a national
basis, also, if the substitution of coal for oil was going well and demonstrably reducing
the nation's oil imports, public attitude towards significantly larger exportation of coal
would be better received than if coal substitution had made no discernible dent in oil
imports.

DEMAND ISSUES

Electric Utilities Sector

The rates of economic growth and therefore of energy consumption and
electricity usage are lower in this study than those in the base case of the National
Energy Plan for 1985 and those in the U.S. submission to IEA for 1990. No U.S. projections of these rates are available beyond 1990. The rates used in this study are displayed below, along with historical rates.

Table I-5 **Average Annual Growth Rates**
Percentage

	1960-76	1976-85	1985-90	1990-2000
GDP	3.5	3.7	2.9	2.2
Energy consumption	3.4	2.8	2.5	1.4
Electricity usage	6.2	4.4	3.4	2.0

Table I-6 **Electricity Sector Inputs in the United States[1]**
Mtce

		1985		1990		2000	
	1976	Low Nuclear	High Nuclear	Low Nuclear	High Nuclear	Low Nuclear	High Nuclear
Electricity production	285.3	421.3	421.3	497.4	497.4	604.9	604.9
Coal	369.6	615.6	615.6	745.7	702.1	800.0	725.7
Oil	120.2	160.3	140.9	146.9	125.0	100.0	70.0
Gas	104.4	57.1	57.1	54.3	54.3	32.9	32.9
Nuclear	66.2	208.6	228.0	292.0	357.5	547.3	651.6
Hydro and Geothermal	95.2	117.7	117.7	139.4	139.4	157.1	157.1
Other	—	4.3	4.3	14.3	14.3	42.9	42.9

1. In this table the nuclear capacities assumed for 1985, 1990 and 2000 respectively are 100 GW, 140 GW and 265 GW
for the low nuclear sub-case and 111 GW, 172 GW and 316 GW for the high nuclear case.

This study also assumes a slower expansion of nuclear capacity than foreseen in
the base case of the NEP. Here it is projected that installed capacity in the low nuclear
subcase will be 100 GW in 1985; 140 GW in 1990; and 265 GW in 2000 (see Table I-
6). This slower projection is assumed to result from delays in completing new generating plants. While such a rate of delay may be anticipated as an average on a national level, it is often unexpected at the specific facility and frequently is accommodated
by adding auxiliary generating units, mostly oil-fired because of their lower capital
cost and shorter pay-out time than those for coal-fired units, until the new nuclear
capacity is operational. For that reason in both 1985 and 1990 oil was substituted for
that small portion of lower nuclear capacity attributed to delayed delivery.

The major substitution of coal for oil as a consequence of slower nuclear expansion first shows up in the projection for 1990; by then much of the new base load coal-fired generating plants will be in operation. By 2000 this substitution could be as great as 74 Mtce (the difference between the lower and higher nuclear sub-cases) as compared to only 44 Mtce 10 years earlier. Higher oil prices by 2000 are assumed not only to spur coal-fired plants but permit the introduction of powerplants fed by geothermal energy and low-BTU synthetic natural gas in a few select regions.

Although the average rate of increase of electricity use of 4.4 per cent through 1985 appears reasonable, it is more than a percentage point below the projections of the Edison Electric Institute. A critical uncertainty surrounds the extent of the possible switch of industry from oil and gas, the use of which would be taxed under the NEP, to either electricity or direct burning of coal for process heat; in either case the demand for coal would rise.

The key to the expansion of greater coal usage in power generation is the cost of meeting the principal environmental problem of burning coal: the emission of sulphur oxides. Coal burning powerplants currently emit one-half of all sulphur oxide emissions. Presently nearly half the coal consumed in powerplants is out of compliance with existing clean air standards. These standards are discussed in Chapter H, and the penalties borne by coal to meet those standards by application of the " best available control technology" are factored into the economics of utility fuels shown in Chapter F.

Programmes since 1973 to accelerate conversion of existing powerplants from oil and gas to coal has shown poor results, while Government estimates of the cost of full conversion have risen from $ 137 million in 1973 to $ 1.2 billion in 1976. The latter estimate would work out according to the General Accounting Office, to $ 63 per kilowatt of capacity or one-eighth to one-sixth of the cost per kw of the construction of a new coal-fired powerplant. As for new powerplants, the industry is planning considerable shift away from gas and oil to coal at the present time but it is not clear to what extent this is a response to the unavailability or higher cost of gas and oil, or to Government orders. The most noted example of conversion now underway is that of gas-fired powerplants of the South Central States, which in 1976 used nearly 70 per cent of all the gas used for power generations in the United States, but are now expected to have completed a total conversion of the base-load capacity to coal or nuclear by 1985.

At this writing, it appears that for the immediate future at least, Congress will withhold from the Federal Government the authority to direct local utility commissions to assist conversion of utilities to coal through mandated load management reform in the form of differential electricity rates for peak and off-peak periods. The initiative now lies with local utility commissions to follow the growing European practice of differential tariffs.

Non-utility Sectors

    a) Residential-Commercial and Transportation.

There appears to be little opportunity to reintroduce direct burning of coal in the U.S. residential-commercial sector, although this sector will nevertheless create a greater derived demand for coal through increased reliance on electricity. This derived demand is already factored into our estimates for coal substitution in the utility sector. The U.S. transportation sector, now reliant upon oil for 96 per cent of its energy needs, has the least flexibility toward shifting to coal or even to highly capital-intensive rail electrification.

b)   Industrial Sector.

A very substantial long-term reduction in the direct-burning of coal in industrial boilers occurred in the United States prior to 1974 as a result of then lower-priced oil and the introduction of stringent air control standards; 20 to 25 per cent of coal-fired boilers were converted to oil in recent years. But since oil and gas prices have increased and the availability of these fuels questioned, the trend away from coal has been reversed and a larger portion of orders for new industrial boilers now specify coal.

Coal Conversion

The Powerplant and Industrial Fuel Use Act of 1978 is designed to encourage the use of coal or other alternative fuel instead of oil or natural gas in both existing and new utility and industrial boilers. The Act prohibits new units from burning oil or natural gas as a primary fuel. Existing powerplants or industrial facilities which are coal-capable may be required to convert to coal or an alternative fuel where DOE finds that it is technically and financially feasible. Existing non-coal capable units may be required to burn mixtures of oil and alternative fuels. However, new and existing units may be granted exemptions from these requirements on the basis of environmental regulations, cost, site limitations, system reliability or the public interest. All existing units would be prohibited from using natural gas after January 1, 1990. The Act provides funds for Federal assistance in various impacts of expanded coal production including regional impacts, the rehabilitation of railroads for coal transport, and to assist firms in purchasing pollution control equipment. The coal conversion measures will reduce U.S. oil imports in 1985 by 0.300 Mbpd.

The Administration's estimates of conversion potential in the industrial sector were based upon a national survey of the age and fuel-burning capability of major fuel-burning establishments. These estimates, however, may have failed to take into account the impact of the Clean Air Act of 1977, the time required for legal clarification of the new environmental regulations, and the present lack of infrastructure for industrial deliveries and ash disposal systems. Nevertheless, economic considerations favour coal, especially in new facilities planned around coal from the beginning, and industrial coal consumption (especially in under-boiler uses) is expected to grow faster than industrial energy demand overall.

CANADA

Canadian coal production, after remaining level throughout the 1960s at about 8 Mtce per year, began to grow rapidly in the 1970s in order to supply coking coal to Japan and low BTU thermal coal to utilities in Alberta and Saskatchewan. For purposes of the present study the most important questions relating to Canada concern: the scope for future coal supply expansion, the extent to which coal can displace oil and gas in domestic consumption, and the prospects for exporting thermal coal.

RESERVES AND SUPPLY

Estimates of coal in place show 28.9 billion tons of Measured Resources, 13.2 billion tons of Indicated Resources and 164.6 billion tons of Inferred Resources. 99 per cent of these resources are located in Western Canada (British Columbia, Alberta and Saskatchewan), and Saskatchewan's share consists entirely of lignite.

These very large numbers do not by themselves help much towards assessing future production prospects. The term " coal resources" refers to concentrations of coal of certain characteristics and occurring in the ground within specified limits of seam thickness and depth from surface. Resource estimates do not show how much coal can be economically recovered.

By means of questionnaires to coal leaseholders th Canadian Government assembled estimates of coal reserves considered economically recoverable at 1976 prices and which meet two criteria:

1. that infrastructure (transportation facilities, electric power, townsite, etc.) either exists or can be amortized from coal sales; and
2. that mining is permitted in the areas by government policy.

The estimates of economically recoverable reserves of thermal coal are summarized in Table I-7.

Table I-7 **Economically Recoverable Coal Reserves in Canada, 1976**

	Mt
Nova Scotia	34
New Brunswick	31
Ontario	NA
Saskatchewan	1 720
Alberta - Plains	1 935
- Outer Foothills	NA
- Inner Foothills	NA
British Columbia - Southeastern	NA
Other	1 010
Canada total	4 729

Source: Department of Energy, Mines and Resources.

These estimates are deficient in that no figures are given for thermal coal reserves in the foothill regions of Alberta, and the southeastern region of British Columbia, which are precisely the areas where future thermal coal exports would likely originate. Those regions, however, are credited with just over half of Canada's Measured Resources.

In 1976, 25.5 Mt of coal were produced in Canada. Of the Maritime provinces production of 2.3 Mt (mostly in Cape Breton Island), 1.3 tons consisted of coking coal consumed locally, in Ontario and abroad, and most of the rest, consisting of bituminous coal, was consumed by thermal power stations in Nova Scotia. Saskatchewan's 4.7 Mt of lignite were consumed largely by mine-mouth thermal power plants as were Alberta's 6.3 Mt of sub-bituminous coal. Alberta's and British Columbia's 11.2 of coking coal were exported to Japan, and 0.8 Mt of bituminous were partly consumed locally and partly exported.

In the absence of detailed estimates of production costs, a rough idea of the latter may be obtained from the following information. Maritime coal is expensive and subsidies are needed to enable it to be sold to Nova Scotia Power Corporation at a price equivalent to that for fuel oil ($ 11.75 per bbl.) The 1976 price at the source of supply for unwashed Saskatchewan lignite was $ 2.79 per ton, while the corresponding price for unwashed Alberta sub-bituminous coal was $ 4.01. The cost of British Columbia bituminous coal delivered to Ontario thermal power stations was about 10 times higher than the delivered price for imported bituminous, which was $ 38.03 in 1976, but transportation charges account for almost 50 per cent of the delivered price. Alberta coal, however, has a sulphur content of only 0.5 per cent compared with 2.7 per cent for U.S. imports. Its attractiveness to Ontario Hydro is such that a new coal terminal was opened at Thunder Bay in September 1978 upgrading a railway line for unit train coal movement between Alberta/British Columbia and the new terminal. Ontario Hydro has contracted with Thunder Bay Terminals Limited to tranship 3.2 mt by 1985.

A study made by the Institute of Energy Economics in Japan contains an estimate that Western Canadian thermal coal could be delivered to Japan at a CIF cost of $37.00 to $40.00 per ton, approximately equal to delivered costs for coal from the United States, and not much higher than delivered costs of Australian coal. Significant quantities of Canadian coking coal are currently being exported to Japan, presumably at prices competitive with those charged by other exporting countries. To conclude, Western Canadian low BTU coal is cheaper than oil or gas for local thermal power generation. Western Canadian bituminous thermal coal is cheap enough to justify sales in Ontario and possibly in export markets as well.

In the past two years Alberta, British Columbia and Saskatchewan have announced new coal policies which will affect future supply, but the exact effects will depend on how the policies are administered.

The three governments plan to permit exports out of their provinces only if it can be demonstrated that a surplus exists after taking account of present and future domestic requirements. Known deposits are sufficiently large that this policy should not impose any constraint in the near future.

British Columbia intends to maintain its present royalty rate of 3.5 per cent of the value for bituminous coal, while Alberta has introduced a formula under which royalties vary from low rates on high cost projects to high rates on low cost projects, and Saskatchewan has announced a royalty of 15 per cent of the value. These three royalty systems do not seem likely to serve as a serious constraint to production.

Both provinces have introduced environmental policies which prohibit coal projects in environmentally sensitive areas, and limit development in certain other areas (for example by prohibiting strip mining). Just how restrictive these policies will be remains to be determined. A mine/power plant complex with an electricity capacity of 2 200 MW proposed by Calgara Power and Canadian Pacific was rejected on the grounds that it would disturb a prime wheat growing area and that alternative sites for power generation could be developed on less productive farmlands.

Another consideration which might affect coal production in Alberta is a possible concern by the provincial government that large new coal projects would contribute to overheating of the provincial economy at a time when substantial oil and gas development is taking place.

New investments by non-residents must undergo review by the Foreign Investment Review Agency, and a number of applications have been approved. It seems unlikely that issues relating to foreign investment will significantly affect coal development.

The projections of future supply assumed in the present study will be discussed after considering demands.

DEMAND

Projections of energy balances for the years 1985, 1990 and 2000 assume annual rates of growth in GDP of 4.1 per cent from 1976 to 1985, 3.4 per cent from 1985 to 1990 and 3.1 per cent from 1990 to 2000.

The coal projections for the low nuclear sub-case are shown in Table I-8.

Demand for coal in thermal power stations needs to be considered in relation to other electricity inputs (see Table I-9). The low nuclear sub-case assumes that nuclear capacity will reach 11.9 GW in 1985, 18.4 GW in 1990 and 36.4 GW in the year 2000. Oil and gas consumption rise between 1975 and 1985, but remain roughly cons-

Table I-8 **Supply, Demand and Trade of Canadian Coal,
Reference, Low Nuclear**

Mtce[1]

	1976	1985	1990	2000
Supply:	20.1	40.1	50.7	71.0
Demand:				
Metallurgical coal	7.8	9.3	10.0	10.6
Electricity generation	14.4	23.7	31.0	45.0
Other uses	1.7	1.1	1.0	1.0
Total	23.9	34.1	42.0	56.6
Trade:				
Net imports (+)/exports (—)				
Metallurgical coal	— 3.8	— 14.6	— 17.3	— 23.0
Thermal coal	7.0	8.6	8.6	8.6
Total	3.2	— 6.0	— 8.7	— 14.4

1. Factors for conversion from tons of coal equivalent to metric tons are: for bituminous thermal and coking coal (1 tce = 1.04 t), for sub-bituminous (1 tce = 1.55 t), and for lignite (1 tce = 2.05 t). Coal exported or imported is assumed to be converted by: 1 tce = 1.04 t.

Table I-9 **Electricity Sector Production and Inputs in Canada
Reference, Low Nuclear**

Mtce[1]

	1976	1985		1990		2000	
		Low Nuclear	High Nuclear	Low Nuclear	High Nuclear	Low Nuclear	High Nuclear
Electricity production	37.0	53.9	53.9	64.4	64.4	90.1	90.1
Coal	14.4	23.7	23.7	31.0	29.1	45.0	39.4
Oil	4.7	9.6	9.6	8.9	8.9	8.9	8.9
Gas	6.3	7.6	7.6	8.0	8.0	8.0	8.0
Nuclear	6.6	24.1	24.1	34.7	38.6	85.5	94.1
Hydro	80.3	101.0	101.0	114.3	112.3	125.4	122.4

1. The nuclear capacities assumed for 1985, 1990 and 2000 respectively are 11.9 GW, 18.4 GW and 36.4 GW under the low nuclear and 11.9 GW, 20.4 GW and 40 GW under the high nuclear sub-cases.

tant thereafter. These fuels would be used largely for intermediate and peak loads and oil may be needed in relatively remote areas where other fuels would be uneconomical. Hydro electric capacity shows substantial growth.

The high nuclear assumption (11.9 GW in 1985, 20.4 GW in 1990 and 40 GW in 2000) results in lower demands for hydro electricity and coal in 1990 and 2000.

All remaining domestic demands for thermal coal are shown to be negligible. This may be a conservative assumption, for several reasons. First, given the relatively low cost of Canadian coal, and the assumption that real oil prices (and hence gas prices which are related to those for oil) will rise after 1985, coal may become competitive with other fuels in the industrial sector (and perhaps in large commercial establishments). Second, coal may be used to make synthetic gas. The government estimates the cost of synthetic gas to be in excess of $ 15 per boe. If it is not now greatly in excess, coal gasification could become economic after 1990. Third, consideration is now being given to using coal to extract heavy oils found along the Alberta-Saskatchewan border. Imperial Oil Ltd. estimates that 25 Mt of coal might be used for this purpose in the year 2000.

Domestic demand for metallurgical coal is assumed to grow at a decreasing rate until 2000.

PROSPECTS FOR SUPPLY AND TRADE

Exports of metallurgical coal are shown to increase from 3.8 Mtce in 1976 to 23 Mtce in the year 2000. Imports of thermal coal (to Ontario Hydro) are assumed to remain constant after 1985.

No exports are shown for thermal coal in the reference case. Given that Canadian coking coal exports can compete in the Japanese market, and given the cost estimates mentioned above, there would appear to be a possibility that thermal coal will find export markets as well. Indeed preparations are currently underway to supply 0.6 mt per year of Western Canadian thermal coal to Japan. Canadian Government officials have estimated that exports of thermal coal from new mines and by-product thermal coal from metallurgical coal operations could reach 9 mt per year by 1985 and could exceed 18 mt per year by 1990 provided that demand develops.

Large scale coal exports would require long advance planning in order to construct the necessary transportation infrastructure, since the coal would have to be transported over a long and mountainous route to the Pacific coast. A possible problem relating to rail transportation is that at present no effective procedures are in operation to regulate unit train freight rates, which would comprise a large part of the cost of delivered Canadian coal.

J

COAL PROSPECTS IN WESTERN EUROPE

UNITED KINGDOM

DOMESTIC SUPPLY

Total established technically recoverable reserves amount to about 45 billion tons of which 6 billion tons are considered economically recoverable. While the latter amount would last for less than 50 years at current rates of production the rate of new discoveries since 1974 has average about 500 mt per annum, which is about four times the rate of production. During 1976 three major prospects were discovered totalling over 800 mt. These discoveries would support a production of about 10 mt per year.

The United Kingdom has experienced a decline in production over the last 20 years from 216 Mt to less than 128 Mt as a result of competition from imported oil. In the aftermath of the oil price increases the National Coal Board prepared plans not just to halt this decline but to bring about an increase in production to 133 Mt in 1985, 142 in 1990 and 167 in 2000. There is, however, considerable doubt whether these targets will be realized.

Existing coal productive capacity has been diminishing at about 3 Mt per year due to depletion of deposits but this decline is expected to fall to 2 Mt per year after 1985 as the industry becomes increasingly concentrated in long life pits. Nevertheless what the actual rate of decrease will be is uncertain.

The Government estimates future production that underlie the high and low growth cases of its Green Paper — "Energy Policy — A Consultative Document":

	Mt		
	1980	1985	1990
Total U.K. coal production[1]	118-128	123-133	112-142

1. Includes estimates of output from projects which have not yet been approved.

Projects currently under construction or merely approved will not be sufficient to meet the NCB's target of 133 Mt by 1985, though the Government argues that more extensive development of existing capacity should enable that target to be met.

No investment strategy has yet been chosen for capacity which would come into production after 1985. There are two additional factors which could constrain development. The difficulty of disposing coal mine wastes together with other environmental problems relating to both production and consumption of coal have generated public resistance to proposals for new coal projects. A second constraint could be a limitation of skilled manpower. A scheme for early retirement was introduced on 1st August 1977 which reduces the retirement age to 60 years by mid-1979, and hence could lead temporarily to a severe loss of manpower. Increases in productivity, however, could permit higher wages making it easier to recruit labour. An incentive scheme was recently introduced which has shown how increased productivity can be linked with higher pay, making it easier to recruit labour. Though output has been declining since 1975 the new incentive scheme seems to have stemmed this trend.

British coal is not as expensive to produce as German, but it is not cheap, and recent estimates suggest that costs will rise, even though new mines are expected to be less costly than current capacity. The average cost in May 1978 was £21.70 (US $40.00) per ton and, on the assumption of a 4 per cent annual increase in wages this cost could escalate in real terms to £24.77 by the late 1980s, although productivity gains may offset the effect of higher wages.

For purposes of assessing prospects for new supply the relevant cost is not the average cost but the incremental cost of new mines. These are expected to be less costly than current capacity but precise estimates are not available.

The reference case projects indigenous coal supply of 111 Mtce in 1985 (127 mt) and 1990, equal to projected domestic demand. Given the expected decline in existing capacity and the limited amount of new capacity under construction the estimate for 1985 may be too high. Supply for the year 2000 is projected to be 120 Mtce (139 mt). The Government's Green Paper accepted the NCB's target of 167 mt for the year 2000 but the Commission assumed that energy prices in that year would correspond to an international oil price of at least double the present level, whereas the present report assumes a level only 50 per cent higher.

DEMAND

The estimates for energy and coal demands (Table J-1 and J-2) up to 1990 are based on official UK projections prepared during 1977, while the projections for 2000 rely heavily on the Green Paper. The differences in the present projections are due to two factors, slower economic growth, and less increase in oil prices. GDP is assumed to grow by 2.7 per cent per year between 1976 and 1985, 2.1 per cent between 1985 and 1990 and 2.2 per cent between 1990 and 2000 (which compares the Government projections averaging about 3 per cent to 2000). Oil prices are assumed to rise in real terms by 50 per cent by the year 2000, in contrast to the Energy Commission's assumption of at least a doubling.

Metallurgical coal demand shows only a very small increase over time.

An increase in demand is shown after 1990 for other uses, which consist largely of demands by industry (apart from iron and steel). The increase is due to a growing gap between oil and coal prices combined with a diminished availability of natural gas in later years.

The Energy Commission projections show a significant use of synthetic gas produced from coal in the year 2000, but this is excluded from the reference case projections given here on the grounds that the cost of synthetic gas will likely exceed the general energy price levels assumed here.

Two nuclear power projections are shown in Table J-2 below. Electricity sector demand for coal is the same in both cases, but oil demand is higher in the low nuclear case. This result is due to the assumptions that domestic coal production would be limited to the amounts shown in the present projections, and that imports of coal would not occur.

Table J-1 **Supply, Demand and Trade of UK Coal**
Reference Case Low Nuclear Assumption
Mtce

	1976	1985	1990	2000
Supply:	103.0	111.0	111.1	120.4
Demand:				
Metallurgical coal	20.3	22.0	22.5	24.0
Electricity generation	62.8	70.7	71.7	74.0
Other uses	18.4	16.9	16.9	22.4
Total	101.5	109.6	111.1	120.4
Trade:				
Net imports (+), exports (—)				
Metallurgical coal	0.3	0	0	0
Thermal coal	0.4	— 1.4	0	0
Total	0.7	— 1.4	0	0

1. The factors for conversion from tons of coal equivalent to metric tons of coal are assumed to be: hard coal for power plants (1 tce = 1.30 t), other (1 tce = 1.05 t).

Table J-2 **Electricity Sector Production and Inputs in the UK**
Mtce

	1976	1985		1990		2000	
		Low Nuclear	High Nuclear	Low Nuclear	High Nuclear	Low Nuclear	High Nuclear
Electricity Production	34.0	40.7	40.7	44.4	44.4	57.1	57.1
Coal	63.4	71.4	71.4	72.4	72.4	74.7	74.7
Oil	17.1	14.4	14.4	14.3	12.1	24.3	14.3
Gas	2.6	2.9	2.9	2.9	2.9	—	—
Nuclear[1]	12.7	19.7	19.7	25.7	27.9	50.0	60.0
Hydro	1.8	1.4	1.4	1.4	1.4	1.4	1.4
Other	—	—	—	—	—	—	—

1. The nuclear capacities assumed for 1985, 1990 and 2000 respectively are 9.3 GW, 11.9 GW and 25 GW for the low nuclear sub-case, and 9.3 GW, 12.9 GW and 30 GW for the high nuclear sub-case.

TRADE

The UK is shown neither to export nor to import significant quantities of coal. The high delivered cost of British coal is expected to preclude exports. On the other hand imports of steam coal are unlikely to supply on a large scale existing power plants because the proximity of many to supplying pits in the UK makes domestic coal competitive with imports. The best opportunity for imports appears to be with new power stations located on the coast.

GERMANY

Germany produces and consumes more coal than any other nation in Western Europe and therefore merits special attention.

DOMESTIC SUPPLY

Technically recoverable reserves are estimated to be 23.9 billion tce of hard coal and 10.5 billion tce of sub-bituminous and lignite, enough to last for 270 years at 1975 rates of consumption. Hard coal reserves which are economically recoverable at current international prices for coal, however, are probably considerably less than 10 billion tce.

Ninety per cent of the lignite deposits are located in the heavily built-up region between Cologne, Dusseldorf and Aachen. Significant expansion of production over present levels is precluded since it would require open cast mines which would necessitate substantial relocation of homes and business activities. The present projections (Table J-3) assume the opening of only one new mine. Consequently lignite production rises from 34 Mtce in 1975 to only 38 Mtce in 2000.

German hard coal is very costly to produce. One study estimates the average cost of German hard coal at DM 163 (U.S. $70) per ton.

For the purpose of comparing the cost to Germany of using domestic coal with the cost of imported energy a more relevant measure would be the incremental cost of domestic coal, that is, the cost per ton of production in the costliest mines currently in operation. The incremental cost would undoubtedly be higher than the average.

Even the average cost, however, exceeds delivered prices of imported steam coal, which range from $33 to $36 per ton, and of coking coal, which range from $50 to $60. While it is not clear from the average cost estimate whether hard thermal coal is cheaper to produce than coking coal, the fact that in early 1976 German thermal power stations were paying nearly $60 per ton for coal suggests that thermal coal costs are not far below the average.

There is little likelihood that the competitiveness of German coal will improve in the future. It is estimated on the assumption of a 4 per cent annual growth in miners' real wages, that the average cost of domestically produced coal will escalate in real terms to DM 173 (U.S. $74) in 1985 and DM 227 (U.S. $97) by 2000. Unless prices of imported fuel increase more rapidly than this, the competitive position of domestic coal will deteriorate. If, however, real coking coal prices rise in the 1980s, as some expect, the competitive position of German coking coal may improve. Coking coal prices need not be closely related to thermal coal or oil prices.

Large subsidies have been necessary to sustain both the demand and supply of domestic coal. According to the "Second Revision of the Energy Programme" (14th December 1977) subsidies relating to coal totalled DM 8.7 billion for the years 1974 through 1977 and they have recently undergone a steep upward trend. In 1977 they were DM 3 billion and in 1978 they are expected to be DM 4.1 billion. Nearly 50 per cent of the 1978 subsidies will be used to promote coal burning in electricity generation, and will be financed by a surcharge of 4.5 per cent on electricity bills. The next largest amount will be used to compensate for the difference between domestic costs for coking coal and the world market price. The remainder consists of socially oriented subsidies to miners and the mining industry. Excluded from these figures are an estimated DM 616 million to be spent in 1978 for coal R & D.

The DM 4.1 billion estimate for 1978 amounts to DM 20 700 per person for the 2 000 000 persons employed in the hard coal-mining industry. Assuming a production of 90 million tons, this is equivalent to DM 45.6 per ton (U.S. $19.50), a large amount when compared with current delivered spot prices for imported steam coal of $33 to $36.

The following two tables contrast the estimates of the Secretariat with those of three independent German institutes on the supply, demand, and trade of German coal to the year 2000.

Table J-3-A **Supply, Demand and Trade of German Coal,**
Reference Low Nuclear Sub-case[1][2]
Secretariat's Estimates

Mtce

	1976	1985	1990	2000
Supply:	126.1	123.6	122.4	125.1
Demand:				
Metallurgical Coal[3]	35.0	32.1	32.0	38.3
Electricity generation	67.1	76.1	91.5	138.7
Other uses	10.0	8.8	8.3	13.3
Total	112.1	117.0	131.8	190.3
Trade:				
Net imports (+)/exports (—)				
Metallurgical coal				
and coke oven coke	—14.9	—13.6	—12.9	—6.9
Thermal coal and products	3.2	7.0	22.3	72.1
Total	—11.7	—6.6	9.4	65.2

Table J-3-B **Supply, Demand and Trade of German Coal,**
Reference Low Nuclear Sub-case [1][2]
German Institutes' Estimates

Mtce

	1976	1985	1990	2000
Supply:	125.9	122	122	126
Demand:				
Metallurgical Coal[3]	35.0	35	54	34
Electricity generation	67.1	68	72	102
Other uses	10.0	9	8	10
Total	112.1	112	134	146
Trade:				
Net imports (+)/exports (—)				
Metallurgical coal				
and coke oven coke	—14.0	—14	—14	—14
Thermal coal and products	3.0	4	6	34
Total	—11.0	—10	—8	20

1. Assumed factors for conversion from metric tons of coal equivalent to metric tons are: hard coal (1 tce = 1 t) lignite (1 tce = 3.7 t) coke oven coke (1 tce = 1.03 t).
2. The difference between domestic supply and domestic demand plus exports in 1976 is accounted for by stock increases and omission of exports of coke.
3. Total domestic consumption of metallurgical coal less exports of coke oven coke.

In 1973 the Government made a commitment to maintain production capacity for hard coal at 94 Mtce to 1980. The " Second Revision of the Energy Programme" does not give explicit numerical targets for 1980 or later years but it is clear that the general thrust of the existing policy is to provide very firm support towards maintaining production levels. To maintain production will be difficult and costly. However, given the magnitude of the subsidies needed to prevent production declines, it is possible that production could be significantly lower than projected here.

DEMAND

Coal demand for electricity generation in Germany must be viewed in context of both projected electricity demand and the availability of fuels for power generation. Table J-4 provides electricity sector estimates under two different assumptions concerning nuclear power. In both cases electricity production grows at the same rate (4.2 per cent per year from 1976 to 2000).

The difference between these two tables is due in part to the difference in electricity generating estimates of the Secretariat and the German research institutes which will be developed below. According to the Government's own estimates the construction time for a nuclear power station is about 70 months, while the total time from application for approval to entry into operation is about 10 years. On the basis of these lead times, those plants which are not already under construction or far advanced towards approval will not be ready by 1985. While it may be possible to place 24 GW of nuclear capacity into operation by the end of 1985 it is doubtful whether this capacity could produce in 1985 the Government's projected production of 156 Twh, which implies an average capacity factor of 74 per cent. In addition the possibility of further delays on environmental grounds cannot be disregarded, though the German Government is endeavouring to clear the authorization of new plants by linking the approval of new construction to the assured availability of intermediate storage of spent fuel in Germany or elsewhere and the operation of new plants to approval of the waste management center.

For these reasons lower nuclear figures were chosen for the present study, 22 GW (120 Twh) under the high nuclear assumption and 18 GW (98 Twh) under the low nuclear assumptions.

The nuclear estimates for 1990 and 2000 were chosen to reflect the possibility of further delays in the nuclear programme, and also to take account of the fact that if non-nuclear capacity must be built to make up for nuclear shortfalls in 1985 and 1990, then the continuing existence of that capacity reduces the need for nuclear power in the year 2000.

The estimates for coal consumption in the electricity sector under both nuclear assumptions reflect the provisions of a recent 10-year contract signed between the electricity industry and the coal industry which guarantee that an annual average of 33 million tons of hard coal will be used in thermal stations up to 1987.

The possible supply gap that would result from delayed expansion of nuclear capacity would likely be filled first by running existing coal-fired power plants (at higher operating rates) in accordance with current German Government planning, and by the construction of new coal-fired power stations. The Secretariat's low nuclear scenario, if realized, would raise the risk that the remainder of the supply gap would be filled by greater use of oil and natural gas by operating at higher rates existing oil and gas-fired plants (which, like coal plants were operating in 1978 at low utilization rates). Ultimately, as the consequence of the lower nuclear scenario, construction of new coal-fired plants would be required, and could lead to hard coal imports as high as 72 Mtce in the year 2000. In contrast, the German institutes' delayed nuclear scenario is one of a far less severe delay and as a consequence the delay is accommodated by both lower electricity production and by the end of the century greater coal usage (including 20 Mtce of imported hard coal).

Table J-4-A   **Electricity Sector Inputs in Germany, Reference Case**
Secretariat's Estimates
Mtce

	1976	1985		1990		2000	
		Low Nuclear	High Nuclear	Low Nuclear	High Nuclear	Low Nuclear	High Nuclear
Electricity Production	41.0	63.9	63.9	80.9	80.9	111.1	111.1
Coal	67.1	76.1	75.9	91.5	89.1	138.7	94.2
Oil	10.6	21.0	17.6	20.1	14.1	15.0	5.0
Gas	17.4	24.3	23.3	24.3	22.1	20.0	10.0
Nuclear*	7.9	30.9	37.8	52.9	71.9	101.9	169.3
Hydro	4.7	7.0	7.0	7.0	7.0	8.0	8.0
Other	3.8	2.4	2.4	5.0	5.0	11.0	11.0

*   The nuclear capacities assumed for 1985, 1990 and 2000 respectively are 18 GW, 28 GW and 50 GW for the low nuclear sub-case and 22 GW, 38 GW and 80 GW for the high nuclear sub-case.

Table J-4-B   **Electricity Sector Inputs in Germany**
German Institutes' Estimates
Mtce

	1976	1985		1990		2000	
		Base	Delayed Nuclear	Base	Delayed Nuclear	Base	Delayed Nuclear
Electricity Production	41.0	64	61	79	74	109	96
Coal	67.1	70	68	75	72	89	102
Oil	10.6	12	11	10	10	3	3
Gas	17.4	22	22	21	20	10	10
Nuclear*	7.9	50	46	83	75	163	118
Hydro	4.7	6	6	7	7	8	8
Other	3.8	7	6	8	7	11	11

*   The nuclear capacities assumed for 1985, 1990 and 2000 respectively are 24 GW, 40 GW and 75 GW for the base case and 22 GW, 36 GW and 55 GW for the delayed nuclear case.

1.   The composition in the electricity sector inputs differ in the two tables largely because of:
   a)   the Secretariat's estimates of nuclear capacity in the high nuclear subcase are lower than the German Institutes' estimates in their base case, and those of the low nuclear sub-case are well below the estimates of the German Institutes' delayed nuclear case;
   b)   the German Institutes assume a slightly lower average economic growth rate (0.4 percentage point lower) in the delayed nuclear case but are nearly equal in the base case with those assumed by the Secretariat.
   c)   Consequently, the assumptions of the German Institutes on electricity demand growth are slightly lower than the Secretariat's.
Generally, the base case of the German Institutes is closer to the high nuclear sub-case of the Secretariat, and the delayed nuclear case with the Secretariat's low nuclear sub-case.

Domestic demand for metallurgical coal is projected to fall below the 1976 level in 1985 and 1990 but rise slightly above it by the year 2000.

Exports of metallurgical coal and coke are assumed to be maintained roughly at current levels until 1990 but fall after that. Since large subsidies are needed to sustain demands for German coking coal, realization of these exports will require continuing and possibly increased support.

The preceding examination of coal demand may be summarized as follows. Both scenarios show domestic coal demands rising above the 1976 level; in the low nuclear sub-case, domestic demand would rise from 108 Mtce to 190 Mtce by 2000.

OTHER EUROPEAN MARKETS

OECD european countries other than Germany and the United Kingdom constitute the largest prospective market for internationally traded steam coal in the future. In 1976 coal consumption in these countries was 125 Mtce of which 57 Mtce was coking coal, 44 Mtce steam coal and 24 Mtce coal used for other purposes. Coal production was less than 56 Mtce in 1976 so as a group these countries were net coal importers of coking coal mostly. Some of these countries once had a flourishing coal mining sector (the Netherlands, Belgium and France) while in others the decline in mining has been reversed with brown coal and lignite deposits showing promise (Spain, Greece, and Turkey).

Poland and the United States in 1976 were the two most important suppliers, the latter mainly of coking coal and the former of both coking and thermal coal. Germany was also a significant supplier of mostly coking coal to her EEC partners.

COAL PROSPECTS

Coking coal demand might remain essentially at the 1976 level through 1985 for all countries included in this section except Turkey where an important increase in steel production has been targeted by the Government. Beyond 1985 modest increases are expected to 2000; however, Turkey should show faster demand growth for coking coal.

In these European countries thermal coal demand does show very high growth potential in the power generation sector and in some industrial sectors (iron and steel, and cement among others), or in energy centres in large industrial parks for combined generation of steam and electricity.

Demand for steam coal in these countries together has shown a reversal of the decline before 1974. Yet, ample coal-fired surplus capacity exists; assuming a 70 per cent load factor for all power stations with coal-burning capabilities it was estimated that potential coal use in electricity generation in 1976 was close to 90 Mtce.

From this base, demand is now expected to grow to 77 Mtce, 107 Mtce and 206 Mtce in 1985, 1990 and 2000, respectively. This implies a 6.6 per cent per annum average growth rate in coal usage or faster than the expected growth in electricity production.

Several countries included in this group merit special attention. In Greece and Spain, a four-fold increase in solid fuel demand for power generation will be satisfied by increases in domestic production — lignite in Greece and lignite and hard coal in Spain. In Italy a ten-fold increase in domestic coal demand is expected to be satisfied for the most part with imports of up to about 22 Mtce in the year 2000. Prospects in Belgium are for a six-fold increase in demand met with imports. In the Netherlands a very substantial increase in coal demand for power generation is likely — to 40 Mtce by the year 2000 in the low nuclear subcase, based on imports. Additionally, coal gazification yielding low BTU gas to be mixed with high BTU gas may be a promising project in the Netherlands.

France, Finland and Denmark do not show very high relative increases in demand but only because they are already the most important steam coal users and importers, in relation to the size of their respective electric systems, in the OECD.

In France, following the 1973-74 oil price rise, the public utility EDF started a definite programme of substituting coal for oil in power generation which, together with a similar programme in Denmark, is the most energetic in OECD Europe. With a domestic high-cost mining industry on the decline, coal imports increased to 11.5 Mt in 1977 from only 2.3 Mt in 1973. South Africa, Poland and Australia have been the three most important suppliers of thermal coal to France. Low shipping costs and favourable FOB prices have recently made the imported coal used by French power plants some of the cheapest used by the electric industry in OECD Europe. In 1977 a new coal terminal was opened in Le Havre which will permit the use of 120 000 dwt carriers in the South Africa-France traffic. The reference case projects steam coal demand for power generation in France to decrease from 19 Mtce in 1976 to 16 Mtce in 1985 (with commissioning of new nuclear power plants and the decommissioning of old coal-fired plants) and increase again up to 43 Mtce in 2000.

Danish utilites have traditionally imported coal on a spot or short-term basis, but their import policy is now shifting towards long-term framework agreements. Poland has been the traditional supplier, but imports have in the last few years shown a pattern of diversification. Thermal coal demand is expected to rise from 3.5 Mtce in 1976 to 6.0 Mtce in 1985, 9.1 Mtce in 1990 and 11.4 Mtce in 2000 in the low nuclear case.

Table J-5 presents the expected coal balances for OECD-European countries excluding the United Kingdom and Germany.

The economic conditions for such a scenario already exist. Since 1975 imported steam coal in Western Europe has maintained with respect to fuel-oil the same price relation (on a calorie basis) which it enjoyed in the early and mid-1950s. Coal has been delivered to power stations with good connections to import harbours in France and Northern Europe at about 15-30 per cent less than fuel-oil. There is reason to believe that European utilities expect that a price differential big enough to justify building new coal-fired capacity (in bivalent or trivalent stations probably) might be maintained in the future.

Two main obstacles to that development are apparent: first, uncertainty regarding government future importing or environmental policies (particularly where siting problems exist due to lack of available land for solid waste disposal, as may be the case of certain congested areas in northern Europe and Italy); secondly, inadequacy of port infrastructure which would have to expand to allow the almost four-fold increase in coal imports foreseen in the projections.

At present very few European ports can receive bulk carriers of over 80 000 dwt. Most coal terminals able to handle these type of vessels serve steel mills and other associated facilities. The limited increase in coking coal imports to 2000 on the European continent can be probably handled with minor expansion of present facilities.

The expected growth in steam coal imports will be borne mostly by sea. A trade expansion of eight-fold magnitude will certainly require new facilities. At present only Rotterdam can service ships of more than 200 000 dwt. This port handles increasing amounts of thermal coal imports destined not only to Dutch consumers but also to French power plants located in the Lorraine (about 1 Mt in 1972) coking coal imports as well as for the British steel industry.

Plans exist for the expansion of Rotterdam Port (Maasvalakte) coal terminal from a throughput capacity of 4 Mt to 20 Mt per year. Rotterdam appears, then, as the first large coal centre in Europe. By 1990 it could use half its throughput capacity for transhipment (via barge or coastal ship) to other European sites. In Denmark work is under way to expand two harbours, Stignaes (to service 125 000 dwt vessels) and Aaberaa (up to 150 000 dwt vessels) which might serve as a coal centre for future Scandinavian imports.

Le Havre, Rouen and Dunkirk are the most important coal receiving ports in France which handled in 1977 almost 13 Mt of hard coal; they could be expanded as well in time to accommodate the larger coal trade expected in 2000. The situation in Southern Europe is, however, less promising. The lack of the canal infrastructure existing in Northern Europe limits to some extent the introduction of coal imports. Nevertheless at least two deep water ports (one in France, and one in Italy) could be turned into coal centres. Investment needed for port revamping is significantly smaller in Northern Europe where the basic infrastructure already exists. In the Italian case lack of adequate sites to build any additional deep water ports makes the use of a regional coal centre even more important.

Coal centres serving large ships will need big stocking areas to permit adequate transhipment. They could be used as coal classification and blending centres which could provide greater flexibility to coal consumers in using different supply sources. This is particularly important as it may diminish the lack of homogeneity which taxes coal (as viewed from the consumers' standpoint) vis-à-vis fuel-oil.

Table J-5   **Coal Balances in OECD European Countries Other than UK and Germany**
Reference, Low Nuclear Sub-case
Mtce

	1976	1985	1990	2000
Supply:	55.6	86.8	98.7	118.5
Demand:				
Metallurgical coal	56 9	62.7	72.2	81.9
Electricity generation	44.3	77.3	107.4	206.3
Other use	24.0	48.5	62.6	75.8
Total	125.2	188.5	242.2	364.0
Trade:				
Net imports (+)/net exports (—)				
Metallurgical coal	42.8	44.3	53.9	63.4
Thermal coal	23.3	57.4	89.6	182.1
Total	66.1	101.7	143.5	245.5

K

COAL PROSPECTS IN JAPAN

Coal consumption in Japan has increased from 70 Mtce in 1965 to 83 Mtce in 1976, an annual rise of 1.6 per cent. This was particularly due to a significant increase of metallurgical coal (24 Mtce to 68 Mtce) which offset the decrease of thermal coal consumption (46 Mtce to 15 Mtce), and accounts for about 80 per cent of present coal consumption in Japan.

Indigenous production has declined from 50 million tons in 1965 to 20 million tons in 1976 mainly in steam coal.

Consequently, coal imports have increased from 18 Mtce in 1965 to 60 Mtce in 1976 to meet total coal consumption. Japanese coal imports are almost totally metallurgical coal and only about 2.5 Mtce of 1976 imports were thermal coal (including coal for gasworks and anthracite).

Geological resources of coal in Japan are estimated to be about 8 billion tons and technically and economically recoverable reserves are estimated at about 1 billion tons. Japanese coal reserves are located in relatively deep areas and their seams are thin in many cases. Therefore, there is no possibility to develop new open cast mines.

COAL PROSPECT

Long term energy supply and demand forecasts were most recently revised by the Japanese Government in June 1977. The recent forecast reflects the assumption of a GNP growth rate of 6.1 per cent per annum for the period 1975 to 1985. Demand for imports of steam coal are forecast at 4.2 Mtoe (6 Mtce) in the "Business as Usual Case" and 11.2 Mtoe (16 Mtce) in the "Accelerated" case for 1985 and at 28 Mtoe (40 Mtce) for the "Accelerated" case in 1990.

The forecast for the future coal market in Japan is also based on the Japanese Government's projection for future nuclear capacity of 33 GW in 1985.

The reference case projection, based on the government projection but assuming lower economic growth and lower nuclear capacity, also suggests a substantial increase of thermal coal import demand in Japan. It is projected to increase from the 1976 level of 1.8 Mtoe (2.5 Mtce) to 9.6 Mtoe (13.7 Mtce) in 1985, 23.2 Mtoe (33.1 Mtce) in 1990 and 53.6 Mtoe (76.5 Mtce) in the year 2000.

ELECTRICITY SECTOR

Reference case projection for coal utilized for electricity generation are 21 Mtoe (29 Mtce) in 1985 and 66 Mtoe (95 Mtce) in 2000 in the low nuclear subcase. It is anticipated that about 80 per cent of consumption will be provided by imported steam coal.

In 1985 and 2000 coal utilized in electricity generation will also include 16 Mtce and 19 Mtce respectively in the low nuclear case, transferred by the use of blast furnace gas and coke oven gas from metallurgical coal. Therefore, 14 Mtce of steam coal

Table K-1   **Supply, Demand and Trade of Coal in Japan**
Reference, Low Nuclear sub-case[1]

Mtce

	1976[2]	1985	1990	2000
Supply:				
Total	20.4	19.0	19.0	19.0
Demand:				
Metallurgical Coal[3]	68.5	92.9	100.0	114.1
Electricity Generation	7.3	13.9	32.1	75.7
Others	7.4	8.6	9.9	10.1
Total	83.2	115.4	142.0	199.9
Trade:				
(Net Imports)				
Metallurgical Coal	57.1	82.7	89.9	104.4
Thermal Coal	2.5	13.7	33.1	76.5
Total	59.6	96.4	123.0	180.9

1.   Estimated by the Secretariat.
2.   The difference between domestic demand and domestic supply plus imports for 1976 is accounded for by stock changes.
3.   Including metallurgical coal which is consumed for electricity generation through by-products of coke oven gas and blast furnace gas.

Table K-2   **Fuel Input for Electricity Generation**
Mtce

	1976	1985		1990		2000	
		Low Nuclear	High Nuclear	Low Nuclear	High Nuclear	Low Nuclear	High Nuclear
Coal[1]	17.3	29.4	29.4	48.8	48.8	94.7	64.3
Oil	100.5	161.7	157.8	163.4	153.9	108.1	101.7
Gas	7.3	47.1	47.1	63.3	63.3	66.7	66.7
Nuclear[1]	11.9	35.1	39.0	65.1	74.6	147.1	183.9
Hydro/Geo	30.9	39.1	39.1	46.9	46.9	55.4	55.4
Total	167.9	312.4		387.5		472.0	
Electricity production	62.9	105.3		132.7		167.4	

1.   Including metallurgical coal products used to produce electricity.
2.   The nuclear capacities assumed for 1985, 1990 and 2000 respectively are 18 GW, 33 GW and 80 GW for the low nuclear sub-case and 20 GW, 37 GW and 100 GW for the high nuclear sub-case.

in 1985 and 76 Mtce in 2000 are required to meet the steam coal demand for electricity generation.

Committed projects for coal-fired electric generating plants are presented in Table K-3. These already committed projects total 5.5 GW of generating capacity and will require 13 Mtce of steam coal per year, part of which will be provided by imports. Thus the Japanese Government forecast in the "Accelerated Case" of 16 Mtce of steam coal import has a good prospect of being achieved. How to reach the government target of 44 Mtce of imported steam coal from 1990 forward remains a problem, short of the conditions assumed in the enlarged coal case.

Table K-3 **Committed Projects of New Coal Power Plants in Japan**
Middle of 1977

	Capacity 10^3 KW	Coal Consumption Mtce	Due Time in Operation
Matsushima (EPDC)[1]	1 000	2 400	1 980-81
Tomato Azuma (Hokkaido Denryoku)	350	840	1 980
Takeharu (EPDC)	700	1 680	1 983
Matsuura (EPDC, Kyushu Denryoku)	3 400	8 160	1 985
Total	5 450	13 080	

1. (EPDC) = Electric Power Development Co.

AVAILABILITY AND COST OF IMPORTED STEAM COAL

Japan currently imports about 2.5 Mtce of thermal coal a year from Australia, USSR and China. Only a token quantity is used for power generation, mainly for testing purposes. A considerable increase in steam coal imports will be required to meet future demand forecasts. Potential suppliers of steam coal to Japan are Australia, U.S.A., Canada, Mozambique, India, Indonesia, Colombia, South Africa and China. In view of the physical coal resources of these potential sources, it should be possible to meet future steam coal import demand requirements in Japan.

The major problem in increasing coal-fired power generation in Japan lies in the willingness of the utilities and the public to accept the costs of constructing such plants rather than the physical availability of steam coal imports.

It is estimated that coal priced at about 40 per cent less than oil, based upon calorific value, would be competitive in the Japanese markets. The Industrial Research Institute of Japan has recently completed a study of future imported steam coal prices for Japan. It is estimated, based on 1976 data, that steam coal imports to Japan would be available in the price range of $30 to $40 per ton CIF.

COAL IMPORT POLICIES

Significant policies being implemented by the Japanese Government to realize the expansion of steam coal utilization are as follows:
— Subsidies to partially compensate the cost differential due to the use of oil and domestic coal.
— Subsidies to the Japan Electric Power Development Company for operating costs of the coal-fired power plant's stack gas desulphurization unit.
— Preferential loans for the construction of newly planned coal-fired plants.

In anticipation of the expansion of steam coal imports the government is proposing to provide the required logistical infrastructure through:
— Subsidies to overseas coal development research projects.
— Financing projects for overseas coal exploration.
— Government guarantees for private loans for overseas coal development.

Public utilities in Japan are also making a concerted effort to use more coal. Utilities have now made commitments for four coal-fired power plants as mentioned before and several others are under consideration. However, utility programmes are hindered by environmental problems and uncertainties in the coal markets. Utilities are requesting that the Government increase the subsidies and guarantees described above, under current Government policies.

COAL CENTRE PROJECT

Japan is studying the feasibility and economic interests of a large coal port which will be capable of accommodating large ocean carriers and with facilities for coal storage — this project has been named "Coal Centre".

The Coal Centre under study is being planned to meet the following conditions:
- — Harbour facilities capable of handling large coal carriers.
- — Location most suitable for economic distribution of coal to consumers.
- — Adequate coal storage facilities.
- — Location with minimal susceptibility to natural disasters such as typhoons, tidal waves and earthquakes.
- — Location most acceptable from an environmental standpoint.

The Coal Centre in vision is planned to handle an annual volume of 7.5 to 15 million tons with storage capacity for 1.25 to 2.5 million tons. It would also be able to accomodate vessels up to 100 000 dwt tons with highly automated loading and unloading equipment. Costs of the Coal Centre are estimated from $125 million for a volume of 7.5 million tons to $250 million for a volume of 15 million tons, or about $17 per ton of coal handled per year in both cases (at 1975 prices).

ENVIRONMENTAL, SITE AND
OTHER PROBABLE CONSTRAINTS

Environmental problems are one of the largest potential constraints to increasing the utilization of steam coal in Japan. The problems of air pollution must be solved through installation of desulphurization and denitro oxygen facilities by coal-fired generating plants.

Equipment for desulphurization is presently in commercial use. It is utilized extensively in oil burning facilities, but presently only to a limited extent in coal burning plants. The main problem encountered in use of desulphurization equipment lies in the amount of dust contained in coal stack gas. This problem is currently handled by reducing the dust with the use of electro filters. Present technology for desulphurization of coal stack gas is about 90-95 per cent efficient — the same as that used in the cases of LNG, naptha, and low sulphur fuel oil.

Technology to remove nitro oxides from coal is still under development and many problems remain to be solved. Costs of this technology are also indeterminate at this time. It can be anticipated that it may require up to 10 years to commercially develop denitro oxygen technology.

Commercialization of the above mentioned anti-pollution technologies is indispensable in consideration of sites for new coal-fired power plants in Japan. The coal power plants have a special problem in requiring space to dispose of fly ash byproducts. Other obstacles for Japan are uncertainties with regard to future coal production policies in potential coal-producing countries and environmental regulations in Japan.

L

COAL PROSPECTS IN OTHER COUNTRIES

AUSTRALIA

Australia is the world's eighth largest hard coal producer. Brown coal and lignite production for domestic use (mostly in power generation) is also important. In 1976, total hard coal production was 68.1 Mt, all of it bituminous coal. Brown coal production was 30.8 Mt.

Hard coal is mined in both underground and open cast operations. In Queensland, almost 87 per cent of all coal produced is currently open cast mined. In New South Wales, however, only 22 per cent of total hard coal production is open cast mined.

In 1976, hard coal consumption was 31.3 Mt and as in other OECD countries, essentially used in coke ovens and for power generation (9.5 Mt and 18 Mt in 1976, respectively). A small and decreasing amount is used in the industrial sector (3.8 Mt in 1976), particularly in cement factories.

Brown coal is produced and consumed in the State of Victoria, almost all of it for power generation.

Overall, some 70 per cent of all power generated in Australia is produced at coal-fired power stations, 20 per cent at hydro-electric stations, and 10 per cent at gas — and oil — fired power stations.

Australia exports roughly half its total production of hard coal (31.1 Mt in 1976). Coal exports from Australia started growing in the early 1960s, spurred by the growth of the Japanese iron and steel industry. More than 75 per cent of all Australia hard coal exports have been to Japan. However, in the recent past, there has been a definite trend towards other markets as well. This is a consequence of both better market conditions elsewhere and a definite policy of export diversification.

Coal is exported from Queenland and New South Wales. All Queensland's coal exports are of coking quality, although some are also suitable in gas works, or the chemical sector. New South Wales' coal exports are for the most part coking coal, although since 1973 steam coal exports have developed. Most of this steam coal was sent to Europe for power generation. The Japanese market for steam coal is not yet developed; deliveries to Japan of steam coal were only 0.3 Mt in 1976.

The long distance separating Australian coalfields from European markets hurts the competitive position of this coal. Furthermore, New South Wales' port infrastructure (the only Australian region to export steam coal) is inadequate for servicing carriers of over 50 000 dwt, thereby denying the transport savings of using larger vessels. Consequently, almost all Australian coking coal exports to Europe have been shipped from Queensland. Hay Point in Queensland has developed a modern coal port and terminal, which can handle 120 000 tonners. The largest coal cargoes known to date have been on the Queensland-Japan route.

Resources and Reserves

According to the latest published estimates, in-situ proved reserves of hard coal in Australia amount to over 36 billion tons of which 20 billion tons would be economically recoverable at present costs and prices. Moreover, this estimate refers only to well known, drilled areas. According to Australian experts inferred coal-in-place resources are over 160 billion tons, of which at least 100 billion tons would be economically recoverable. About one-fourth of the proved reserves could be open-cast mined.

Brown coal reserves are also large: 67 billion tons are now proved, of which some 31 billion tons could be economically recoverable.

Criteria used to classify hard coal economically recoverable reserves include a maximum depth of 600 metres, minimum seam thickness of 1.5 metres, and maximum ash content of 30 per cent (dry basis). For brown coal maximum depth is 300 metres and overburden thickness of up to 90 metres; the minimum seam thickness is 3 metres.

All types of hard coal exist in Australia but for the most part it is coal with a high volatile content and ash. Coking coal almost always requires some washing to bring the ash content down to commercial limits. Some 60 per cent of Australian hard coal proved reserves are of coking quality. Steam coals have a high ash content but also have a high ash fusion temperature. For the most part, they are very low in sulphur (less than 0.8 per cent in weight).

The present level of Australian proved, recoverable coal reserves could sustain a much higher lever of production. At 1976 production levels, proved reserves would last more than 200 years.

Prospects for Coal

Coal production costs vary widely according to the type of operations and the structure of the seams. The Australian coal industry is one of the world's most efficient, measured in terms of labour productivity.

Capital costs for coal mine operations vary widely. A minimum would be US $ 25 per ton per year in some of the most favourable extensions of the known New South Wales coal areas. For completely new production in Queensland, however, capital inveestment could reach U.S. $ 120 a ton per year (late 1976 U.S. dollars). This high figure includes the cost of developing the transport infrastructure and of establishing new towns for workers and their families in a sparsely populated area.

In late 1976, total hard coal mining costs for new underground operations in Australia could be estimated at some $ 18 per ton, of which 50 per cent would represent labour costs. Total coal mining costs for new open-cas development in Queensland are lower (from some U.S. $ 8 per ton for the most favourable cases up to $ 15 per ton in the higher cost areas).

Transportation costs from mine to export port are surprisingly high considering that most measured coal reserves are located within 300 km of the coast. Railways are operated by public companies under the control of State Governments. Rail transport costs may be as high as U.S. cents 1.7 per km for a distance of over 200 km and even higher for shorter routes.

Australian port charges are also quite high when compared to other hard coal loading areas of the world. In New South Wales, terminal charges are some U.S. $ 4.5 per ton, perhaps because of over capacity in terminal design.

In 1975, the Australian Government introduced a tax on all coal exported from Australia ($ 6 per ton for hard coking coal and $ 2 for all other) but the tax was soon phased out on steam coal exports.

Table L-1 shows Australian hard coal projections to the year 2000. Brown coal is included in this Table. In 2000, brown coal production could be as high as 70 Mt per year, most of it for use in power generation in Victoria. Hard coal production is projected to be 105 Mt in 1985 and 260 Mt in 2000. The levels of production could be attained with growth rates lower than those observed in the last ten years.

By 1985, hard coal exports could be almost double those of 1975, or 57 Mtce. It is expected that some 15 Mt will be steam coal exports mainly to Japan but also to Western Europe and other Asian countries (e.g. Taiwan, South Korea). By 2000, internal coal demand could be some 90 Mtce and exports could be 195 Mtce. In the 1990s, the volume of steam coal exports should overtake that of coking coal exports.

This export level assumes Australian coal production can expand to 285 Mtce without a large increase in real cost of coal mining. Australian coal supply curves are presumed to be rather flat, for coking and steam coal open-cast mines in Queensland and New South Wales.

In time, it should be possible for Australian coal to maintain a favourable competitive position in the Japanese market and other fast growing Asian markets against residual oil and even other sources of coal.

There do remain, however, some problems regarding capital availability, government surveillance of coal export contracts and infrastructure development.

The Australian Government aims at a level of 50 per cent of Australian held equity in all new mining ventures (75 per cent for uranium), including coal mining. If this policy is maintained the availability of investment capital could limit the projected four-fold increase in coal production in Table L-1. However, it seems that some flexibility may be incorporated in the practical application of the policy guidelines. Government policies do not seem to have discouraged some international oil companies from recent acquiring coal leases and mining operations, with a view to the export potential for both coking and steam coal.

At present, all export contracts have to be reviewed by the Commonwealth Government before issuance of an export permit. The Government may refuse an export permit if prices are below " market prices". While market price is observable in the established world market for coking coal, it is not in the undeveloped international steam coal market. With maritime transport costs accounting for a singificant proportion of CIF price of potential Australian coal deliveries, specification of " market price" in either FOB or in CIF terms could make a considerable difference. Long-term contracts will be absolutely necessary to guarantee new production capacity, particularly for Queensland coal mines with high capital-intensiveness. The Australian Government is promoting a long-term contract for steam coal which is similar to that of coking coal. Coking coal contracts are supply agreements, subject to annual or bi-annual price negotiations, with delivery price escalating the base price between periods of negotiation. If the parties fail to agree after a set time, the contract terminates. For the emerging steam coal market, the Australian Government has suggested introduction of this type of contract renegotiable every three to four years with full escalation in between (including any increases in taxes and royalties). Foreign utilities appear rather reluctant to accept an agreement of this character covering sizeable volumes.

Another more serious obstacle to increased coal production for export is the lack of adequate infrastructure. By 2000, the railway network will have to be revamped if greatly increased coal volumes are to be transported. Railway companies are owned by State governments. As was mentioned earlier their present coal tariff rates are high. An associated obstacle to coal exports is port infrastructure. Presently some 60 Mt per year throughput capacity exists in Australian coal ports. In New South Wales ports, from which all present export steam coal cargoes are sent, maximum capacity is limited to 50 000 dwt. In order for Australian coal were to realize its full export potential to distant markets it would have to be carried in the largest vessels available which would be ineffective. Thus a significant port and terminal development programme in Australia appears in order.

Through 1985, port capacity limitations could seriously impair the competitive position of Australian steam coal in European markets vis-à-vis South African and Polish coal. Nonetheless, starting in the late 1980s, with completion of infrastructural expansion, Australian coal, low in sulphur content, could win new major export markets. By the end of the century Australian steam coal might have a dominant position in the world.

Table L-1   **Australian Coal Balance**

Mtce

	1976	1985	1990	2000
Production	69.6	108.9	151	285
Exports	31.0	57	90	195
of which				
— Coking	27.9	43	54	75
— Thermal	3.1	14	36	120
Consumption	37.3	52	61	90
of which				
— Coke ovens	9.3	12	14	17
— Electricity generation	24.8	38	44	70
— Other	3.2	2	3	3

POLAND

Poland's energy economy is characterized by abundant coal deposits but scarce resources of other energy forms. Poland receives most of its oil and gas supplies from the Soviet Union at prices that advance according to a 1974 schedule and will approach wolrd prices by 1979. These deliveries from the USSR are not likely to be expanded. In order to meet its growing domestic energy needs, Polan will have to further exploit coal production and curb export expansion, while at the same time importing more oil from sources other than USSR.

Poland possesses extensive, developed hard coal deposits which are favourably located, and brown coal deposits which are not yet fully exploited. Besides these energy reserves, Poland has only limited quantities of natural gas with a high nitrogen content, more valuable for fertilizer than energy production; and only negligible oil, uranium, and hydro resources.

Almost 85 per cent of total energy requirements are satisfied by solid fuels (hard and brown coal). Hard coal production in 1976 was some 180 Mt of which 38.9 Mt (22 per cent of total production) were exported. Exports to other Centrally Planned Economies amounted to 13.5 Mt. Out of the 25.4 Mt exported to non-CPEs, 15.1 Mt have been estimated by the Secretariat to be thermal coal exported mainly to Western Europe where for the most part it was used in electricity generation, particularly in France, Finland and Denmark.

Resources and Reserves

Hard coal: economic reserves are located in the three regions containing coal seams: Upper Silesia, Lower Silesia and the Lublin areas, the first two of which are producing areas and the latter under development. In all they total 57 billion tons (in January 1974). Table L-2 below gives a breakdown of the hard coal reserves according to their stages of development:

Table L-2 **Polish Hard Coal Reserves[1] (1st January, 1974)**

	Hard Coal		Of which coking coal	
	Billion tons	per cent	Billion tons	per cent
Total	57	100	13	100
ol which :				
Existing Mines	25	43.6	5.5	42.3
Mines under Construction	1.7	3.0	.7	5.4
Non-developed Deposits	30.5	53.4	6.8	52.3

1. Criterion retained: depth of exploitation not below 1 000 m, thichness of seam minimum 80 cm for steam coal and 60 cm for coking coal, dip maximum to 20°.

Of the 57 billion tons of hard coal economic reserves, about 55 billion occur in the Upper Silesian coalfield, 0.5 billion in the Lower Silesian Coalfield, and about 1.6 billion in the surveyed region of the Lublin coalfield. Total prospective hard coal resources of Poland are estimated at about 70 billion tons (of which about 30 billion occur in the Upper Silesian and 40 billion in the Lublin regions).

The general characteristics of Polish coal are: low ash content; good grindability; and low sulphur content (about 1 per cent).

Brown coal: brown coal reserves are widely dispersed over Poland but the three major fields are located in the central and south west areas: the aggregate of proven and recorded reserves amounts to 14 billion tons. The depth of seams is from tenths of metres to more than 200 m, and the ratio of overburden to coal varies from 2.6 to 9.0. As to quality, Polish brown coal contains about 50 per cent of moisture and has calorific value varying from 1 600 to 2 400 kcal/kg.

Coal Prospects

Total primary energy needs of Poland are expected to continue to increase at an annual rate of about 4 per cent, creating favourable market conditions for conventional coal demand and new prospects for the chemical processing of coal.

Rapid expansion of the coal mining industry is required for many years ahead in order to support national economic development. The possibility of stagnant oil and gas imports from the Soviet Union at subsidized prices, the increase in internal energy demand and rising debt position vis-à-vis OECD countries will force Poland to expand internal coal production to substitute for oil imports and get as much hard currency as possible. The current five-year economic plan as well as the prospects to 1990 call for annual increases of 6 to 8 million tons in hard coal production. Specifically, this rate, if sustained, would lead to the production ranges shown in Table L-3.

Table L-3 **Projected Coal Production in Poland**

	Million tons	
	Hard Coal	Lignite
1980	200	40-45
1985	230-235	85
1990	250-265	104-110
2000	300	150-200

The national plans anticipate these demand trends:

— Increased electricity demand for steam coal, which is currently met mostly by hard coal and lignite (respectively 63 and 30 per cent). Expanded solid fuel fired generation is currently under construction or decided which would bring the total installed capacity of 51 000 MW by 1985 from the 1976 level of 21 000 MW. About 5 000 MW new capacity will be fired with lignite extracted in two new fields;

— Demand for coking coal from a new large iron and steel complex in South Poland;

— A new demand for large quantities of brown coal and hard coal for chemical conversion of coal into gas of high calorific value, liquid fuels and solid smokeless fuels. To promote these processes, conversion of hard coal into manufactured gas might occur in the early 1980s.

The coal industry seems to have encountered serious difficulties to meet expected demand and power shortages have been recently evident in different parts of the country due to coal unavailability.

Production bottlenecks can be expected to continue in the mid-term. The consequence could be a levelling off of hard coal exports through 1990. New nuclear power stations to come on-stream in the early 90s could release additional quantities of stream coal exports to 2000.

In summary, it may be said that plans for the expanded production of hard coal (coking and steam coal) will support export volume at the present level until at least 1990. The Polish authorities are, however, increasingly insisting on financial participation by the countries which intend to import Polish coal, to finance the construction of additional production capacity and the necessary infrastructure. This would promote a shift towards longer-term contracts. Nonetheless, Poland cannot be looked to for significant new volumes of coal to sustain enlarged worldwide trade.

Table L-4 presents the Secretariat estimate for Polish exports of hard coal to non-Centrally Planned Economies.

Table L-4   **Poland: Estimated Hard Coal Exports to Non-CPEs**

Mtce

	1976	1985	1990	2000
Tota Exports	23.9	25	25	30
of which				
Coking	10.3	10	10	10
Thermal	13.6	15	15	20

OTHER CENTRALLY PLANNED ECONOMIES

The Centrally Planned Economies (CPEs) of Europe and Asia are important producers and consumers of both hard coal and lignite. In 1978 the CPEs produced roughly a third of all commercial primary energy produced worldwide, but their share of solid fuel output was over 50 per cent of the world's total.

In addition to Poland, whose prospects have been discussed above, both the Soviet Union and China merit some attention in trying to assess their future impacts on the international steam coal market.

USSR

Coal production in the USSR was some 705 Mt in 1976 (495 Mt of hard coal and 210 Mt of brown coal, lignite and peat). Almost 70 per cent of all coal was produced in underground operations (compared to 75 per cent ten years earlier). The rate of growth in coal production has slowed from 5 per cent yearly in the late fifties, to 2.5 per cent yearly in the early 70s and to only 1.5 per cent since 1975, as a result of the shift to oil and natural gas in energy production.

Solid fuel share of total primary energy requirements has been steadily decreasing from almost 50 per cent in 1960 to some 30 per cent in 1976. Out of a total coal consumption of 500 Mtce, 200 Mtce were used in electricity generation, some 120 Mtce in coking, and 180 Mtce for industrial or other uses.

Coal exports in 1976 totalled some 27 Mt but only 11.9 Mt were shipped outside the Eastern European area. Imports into OECD countries of USSR coal in 1976 amounted to about 9.2 Mt to OECD (of which 4.8 Mt were coking coal).

CHINA

China is the third largest coal producing country in the world but the dominance of coal in the domestic energy economy has shrunk. Coal production (all grades) in 1976 has been estimated at some 470 Mt, some 65 per cent of total primary commercial energy production, down from almost 98 per cent in 1952, as a consequence of the discovery of large oil fields. Coal output has been growing at about 6.0 per cent yearly but has failed to keep abreast with domestic demand. The somewhat obsolent industry received a serious setback by the severe earthquake in the Peking region in 1977.

There has been little exportable surplus of coal; no more than a total of 3 Mt per year of hard coal has been exported from China to Japan and other Asian countries.

Resources and Reserves

The Soviet Union possesses ample reserves of coal of all types. The World Energy Conference (WEC) survey reported 83 billion tce of economically recoverable USSR coal reserves and 27 billion tce of brown coal. Some 17 billion tons may be of coking quality. Published Soviet sources fix proved plus possible reserves of coal-in-place at 256 billion tons of all types. Some 90 per cent of the reserves of coal are located east of the Urals, far from the most important consumption centres of the USSR.

Coal reserves in China are also important. The WEC puts them at some 100 billion tce for hard coal, of which some 20-25 per cent would be of coking quality.

Coal Prospects

In the wake of higher energy prices, Soviet planners are trying to reverse the long-term trend away from coal. Particular emphasis has been placed on greater coal use for electricity generation, with a target of holding coal's share of fuels for power generation to 43 per cent by 1980, which would be still lower than the level of 1970 (some 46 per cent).

The long-range target for coal production appears to be to reach an annual growth rate of some 2.5 per cent through 1990 to reach then a production level of 1 billion tons. The shift in production from the traditional European USSR area to the important coal basins east of the Urals will certainly pose serious problems, particularly related to timely development of transport infrastructure. Already lack of availability of rail cars along with production capacity shortages are making very difficult the attainment of the current 5-year Plan (805 Mt by 1980) placing pressure on the fuel-oil supply for power generation.

On the international trade side some limited expansion of present export levels can be expected from the Soviet Union. The traditionally exporting areas of the European USSR might continue to supply coal to non-CPE countries at most to the present levels, which have not been changing since the late 60s. Furthermore, exports of coal (both coking and steam) from Eastern Siberia could grow in the mid-term. A new deep water port, under construction at Vostochniy on the Siberian Pacific coast, will handle 40 Mt throughput of coal from the Southern Yakutian basin, 1 200 km to the West. Much of the coking coal will be exported to Japan starting in the early 1980s under a long-term supply contract.

Prospects for coal in China are difficult to establish. The current drive to modernize the economy will intensify future energy needs. But the lack of adequate resources allocated to the coal mining industry and the favoring of a faster development of oil will tax coal authorities in meeting domestic demand and might preclude any important development of exports, in the Kiangsu Province. The new coal port in Lienyunkiang seems to imply, however, a moderate increase of coal exports in the future. The Secretariat's tentative estimates for coal exports from China and the USSR (net of exports to other CPEs) are presented in Table L-5. There is great uncertainty about these figures, particularly for Chinese coal exports.

Table L-5   **USSR and China: Projected Coal Exports to the Non-CPE Countries**
Mtce

	1976	1985	1990	2000
USSR	11.9	15	20	30
of which :				
— coking	7.5	10	13	15
— thermal	4.4	5	7	15
China	0.3	3	4	6

SOUTH AFRICA

South Africa is probably the only developed market economy primarily fuelled by coal. In 1976, coal supplied almost 75 per cent of total primary energy requirements with oil supplying 25 per cent. Coal consumption in 1976 reached some 71 Mt, most of it for power generation but also for direct use in industrial transportation and domestic residential sectors as well as for conversion to liquid and gaseous fuels. South Africa is the only country where commercial coal liquefaction is carried out.

Both bituminous coal and anthracite are produced in South Africa, the latter mainly in the Natal region and the former in the Transvaal and Orange Free State. Total coal production was over 76 Mt in 1976. Operations are mostly underground (85 per cent of total coal mined) but the share of open cast mining has been increasing steadily as labour costs grow. Undergound operations in South African coal mining employ relatively little capital in relation to labor, resulting in comparatively low rates of coal-in-place recovery.

South Africa has traditionally exported small quantities of coal. Until 1973, total annual exports were about 1.5 Mt of which more than half was anthracite. The main customers were the CEE countries, particularly France, where South African anthracitic coal was used mainly in the iron and steel industries. Japan also imported limited quantities of bituminous steam and soft coking coal from South Africa.

However, since 1973 South African coal exports have changed remarkably. Coal exports doubled between 1975 and 1976, rising to 6.0 Mt, making South Africa the 7th largest coal exporter in the world. This expansion was made possible by the opening of a new coal terminal at Richards Bay. This port, one of the world's most modern, presently has a capacity of 12 Mt per year which will be increased to 20 Mt by 1978. At present, the port can service 150 000 dwt vessels at a loading rate of 6 500 tph; ultimately 250 000 dwt vessels could be handled. Richards Bay is connected to the Transvaal coal producing region by a railway line built exclusively to service the port.

Once again in 1977, exports doubled from the previous year, promoting South Africa to the 5th largest coal exporter. Some 2.2 Mt of anthracitic and 10.5 Mt bituminous coal, almost all thermal coal, was exported. The most important market was the CEE region which took 7.8 Mt of South African hard coal in 1977.

The internal coal market is closely regulated in South Africa with mine-mouth prices fixed by government. Internal production and exports are allocated to individual collieries by producers' associations. Recently, some international oil companies started operating in South African coal and began exporting steam coal. As of 1978 export permits for some 20 Mt of coal per year have been granted.

Resources and Reserves

The publication in 1975 of the Report of the Commission of Inquiry into the Coal Resources of South Africa, known as the Petrick Report, provided the most updated estimate of resources and recoverable reserves. Table L-6 reproduces a summary of reserve estimates. Only coal deposits situated no lower than 300 meter depth and seam thicknesses of over 1.2 meter or over (0;7 meter for coking coal and anthracite) were considered.

Proven, indicated and inferred coal reserves were established in the Petrick Report at some 81 billion tons. Only 25 billion tons are considered to be exploitable in underground operations; this estimate is very low because the highly regulated market has led to a low average capital intensiveness in mining and a low average rate of coal-in-place recovery. Since the report was published, three new coal fields have been located but their reserves are not known. With present prices for coal in the international market, the Secretariat estimates that economically recoverable coal reserves in South Africa could be as high as 55 billion tons.

Table L-6 **South Africa's Hard Coal Reserves[1]**

Billion tons

Coal-in-place[2]	82.0
of which :	
Underground mineable[3],[4]	25.3
Open-cast mineable[3]	27.8

1. Proven, indicated and inferred reserves.
2. Raw coal (non-washed) of up to 36% ash content. Depth limit of seams: 300 m. (500 m. for coking coal and anthracite). Seam thickness: over 1.2 m. (over 0.7 m. for coking coal and anthracite).
3. They cannot be added to give total extractable coal since there is overlapping in the figures.
4. On the basis of traditionally low recovery factors using mainly the board and pillar mining technique.

Hard coal reserves in South Africa are mostly composed of bituminous steam coal, although limited quantities of anthracite and coking coal exist. The typical heat content of South African bituminous coal is not very high, even when washed. High grade steam coal ranges between 6 300-6 600 kcal/kg, although some coal being exported may be as low as 6 000 kcal/kg. Ash content, as mined, varies between 7 per cent and 30 per cent. However, South African coals tend to have high ash fusion temperatures. Sulphur content tends to be low. An average of 1 per cent in weight would be a representative value. Some 50 per cent of total reserves of saleable coal would be open-cast mineable.

Coal Prospects

South Africa's energy policy has always been remarkable for its high reliance on domestic coal. The most recent indications of policy point to even more intensive use of coal in the future. Also liquefaction of coal is being enlarged by a second larger oil-from-coal plant, despite reportedly unfavourable economics.

Coal costs in South Africa tend to be low despite the predominance of underground operations and high labour intensiveness. Some U.S. $ 15 (1976 dollars) per ton for undergound mined coal would be a representative maximum coal cost at mine mouth. For strip mining operations a representative cost for washed coal could be U.S. $ 8-10 per ton. Capital costs per ton/year of production are also low, averaging about $ 28 per ton/year ($ 40-45 for the U.S. coal mining industry), including the coal washing plant (1976 dollars).

Labour costs, though still low in comparison to those in OECD producing countries, have increased since 1974 much faster than the general price index. Mining costs in South Africa, moreover, are not expected to increase in real terms in the mid-term, because ample opportunity exists for employing more capital equipment per worker. These circumstances should permit to maintain a favourable competitive position for South African steam coal in the European and Japanese markets, assisted in part by the low sulphur content of South African coal. Also, rail transport costs from the main producing regions to export harbours are to remain constant in real terms through the year 2000 at about 1 cent per ton/km.

Despite favourable commercial prospects for exports, government export policy is unclear. In South Africa there exists a nationalistic school of thought advocating more reliance on coal for security reasons, banning exports and reserving coal for the domestic market, including use in oil-from-coal plants. By contrast coal producers argue for further development of coal exports. So far the Government has not defined any long-term strategy on coal trade, but it does not seem to put up any barriers to the promising start in steam coal export trade. Table L-7 shows a list of on-going coal projects. Among them some 10 Mt per year are earmarked for exports by the early 1980s.

Coal prospects for South Africa (and in general the energy outlook) have been studied by the South African Department of Planning and the Environment. The following analysis is based on 1977 data.

Two different economic growth scenarios have been identified for the purpose of energy forecasting and are defined in Table L-8.

In the 1960-73 period GDP annual growth was 5.5 per cent whereas OECD's growth was 4.4 per cent. If the same trend were to continue, only the lower growth scenario above would be compatible with the OECD growth scenarios of this report.

On the basis of the economic growth rates above, two different sets of coal demands have been developed by the Department of Planning and the Environment; these are displayed in Table L-9. A definite policy to rely on coal for future energy needs is implied throughout the study.

Table L-7   **New Coal Mining Projects in South Africa**

Mine	Company	Output million ton/year	Use	Starting Date	Type
Bosjesspruit	SASOL	10	Conversion	1981	Underground
Duva	Barlow Rand (Witbank Co. Ltd.)	9.5	Escom	1979	Surface
Ermelo	GM, BP, Total	3	Export	1979/80	Underground
Grootgeluk	Iscor	1.8 2	Coking Steam	1982	Surface
Kleinkopje	AAC	2 1.6 0.7	Export Local Low ash blend coking coal	1979	Surface
Kriel	AAC	8.5	Escom	1978	Surface
Matla	GM	9.6	Escom	1980	Underground
Rietspruit	Barlow Rand (TCL), Shell	5	Export	1979	Surface

Source: Chamber of Mines of South Africa.

Table L-8   **South African Economic Growth Scenarios**
Per cent per annum

	1974-80	1980-2000
High	6.4	5.5
Low	5.5	4.7

Table L-9   **South African Coal Balance**
Mt

	1976	1985 H.G.	1985 L.G.	1990 H.G.	1990 L.G.	2000 H.G.	2000 L.G.
Production	76.4	137	130	189	177	292	262
Exports	6.0	38	38	67	67	100	100
Imports	—0.3						
Consumption	71.0	99	92	122	110	192	162
of which :							
Coke Ovens	9.2	10	9	13	11	25	20
Electricity Production	41.2	53	49	69	62	118	101
Other Uses	20.6	36	34	40	37	49	41
Memorandum item							
Exports in Mtce	5.6	34	34	60	60	90	90

H.G.   High Growth case.
L.G.   Low Growth case.

Despite the recent discussions about an integral utilization of coal (the COALCOM the concept − Coal, Oil and Megawatts) which would set up coal-fuelled integral energy parks, coal demand forecasts above do not contemplate oil-from-coal plants other than those already operating or under construction, (SASOL I and II, respectively). The increase in coal use will largely depend on the increase of electricity in the overall final energy demand mix. Electricity may increase from the present 17 per cent of final demand to 28 per cent in the year 2000. The steel industry would continue to be the second most important outlet for coal.

There appears to be no resource constraint upon steam coal exports up to the year 2000. Presently exploitable coal resources are ample and higher export prices would permit a higher rate of recovery. Coal exports should not create any scarcity of coal for the domestic market.

The infrastructure for coal development does not represent a potential constraint. In a very short time coal export capacity may rise to over 40 Mt per year and the success of the Richards Bay development (including building the railway connections) indicates that infrastructure bottlenecks are not likely. Furthermore, the terrain between the Transvaal coal fields and the possible port sites on the Indian Ocean coast is favourable to railroad construction. The port of Maputo, Mozambique, could be slightly modified to take a large annual throughput if conditions were appropriate.

Coal export levels will essentially be a matter of policy. It has already been mentioned that a segment of South Africa's public opinion favours a rather stringent conservation policy which would reserve internal energy resources for future energy needs. Also, coal owners are obviously eager to profit from the coal bonanza if it ever develops.

In Table L-9 coal production and exports to 2000 have been estimated, both for the low and high growth cases displayed in Table L-7. The only distinction between the cases has to do with the rate of increase in demand in the domestic market. The projected export demand remains constant on the premise that since the demand is foreseen, the price of coal is set by government, and not important long-term production constraints are present, no reduction in export market in the long-term will be required to satisfy domestic demand should the higher growth rate be realized. The assumed levels of production and exports imply an annual growth rate of production of over 8 per cent through 2000, significantly higher than that observed in the past. This could require a reallocation of resources within the mining sector in South Africa which the expected world market conditions would justify. As shown in Annex 1, Table 1-3, South African coal might play the dominant role in the emerging steam coal market through the early 1990s despite some likely short-term growth problems discussed in Annex 2.

Should South Africa choose to go to further vertical integration in industry in order to export products with higher value added content, especially into ferro-alloys and more particularly into ferro-chrome, the energy-intensiveness of these industries will increase domestic demand for steam coal to generate the required increments of electric power. This possible development is not taken into account in the projected domestic consumption; however, no significant long-term production constraint is foreseen that in itself would require a restriction of steam coal exports in order to satisfy intensified industrial development.

DEVELOPING COUNTRIES

Except for the case of India and South Korea, solid fuels are not very important components of the energy balances of the less developed countries (LDC) − on the average 18 per cent of their total commercial energy requirements.

Hard coal production in all LDCs was about 140 Mt (some 6 per cent of world's total production) in 1976, of which 100 occured in India and 16.4 Mt in South Korea. Some 9 Mt of brown coal and lignite were also produced in LDCs in the same year.

Resources and Reserves

Little is known of actual coal resources in LDCs; except in India and South Korea, little exploration was carried out prior to 1974. The survey of the World Energy Conference indicated a total of economically and technically recoverable reserves of some 55 billion tons of coal equivalent. It is not known, however, to what extent this estimate refers only to known coal bearing areas and to what extent it includes extrapolation to non-explored regions. Half of these estimated reserves are located in the Indian Sub-Continent, some 35 per cent in Latin America, some 17 per cent in African LDCs and the rest in other Asian countries.

It appears that nearly all these coal reserves are of thermal rather than coking quality. The prevalence of coal could permit a relatively rapid growth of coal consumption in many no-noil LDCs. However, not all the coal, because of metal traces or other undesirable qualities could compete in international trade.

Coal Prospects

Despite the difficulties involved in making overall energy projections for developing countries, especially for coal where data are so incomplete, an attempt was made by the Secretariat to assess possible future coal production, usage and trade by LDCs. The results are presented as global aggregates in Table L-10.

Table L-10   **Hard Coal Balances in Developing Countries**
Mtce

	1976	1985	1990	2000
Production	122	177	240	404
Net imports (+)/exports (—)	8	11	2	—10
of which :				
— Coking	8	16	19	26
— Thermal	—	—5	—17	—36
Consumption	130	188	240	390
of which :				
— Coking	34	49	57	76
— Thermal	96	139	183	314

In 1976 hard coal consumption in LDCs was 130 Mtce, of which 34 Mtce were coking coal and the rest thermal coal (both bituminous and anthracitic). The relative share of India and South Korea in the overall thermal coal demand was probably close to 95 per cent.

Coking coal consumption was more divided among LDCs with India, South Korea and Taiwan in Asia and Brazil, Mexico, Argentina and Colombia in Latin America, as the most important consumers.

A total of 8 Mtce of coking coal represented nearly all the net coal imports into LDCs, mostly to Brazil and other Latin-American countries. Some exports from India and Mozambique were also recorded; Indian exports appear to be thermal coal. Exports did not reach 1 Mtce in 1976 although marketing activities in the European and Japanese market to increase the level of exports was evident.

Consumption of both coking and thermal coal is expected to increase rapidly, to an estimated 49 Mtce and 139 Mtce, respectively, in 1985. Coal consumption will thus increase at an average rate of 4.2 per cent in the 1976-85 period, lower than that expected for total energy requirements (TER). In the 1985-2000 period, however, growth in coal consumption in LDCs could be 5.0 per cent yearly or approximately equal to the expected growth in TER. However, thermal coal demand should grow at a faster rate (say 5.6 per cent) than overall coal demand after 1985 as a consequence of the demand for coal for electricity generation.

The relative scarcity of coking coal reserves in LDCs coupled with an expected rapid growth in LDCs steel production (as the countries build basic industries) will produce a further increase in coking coal imports that may double in 1985 to 16 Mtce. However, good prospects for steam coal production may make the LDCs (taken as a group) net coal exporters if the markets for it develop. Steam coal net exports should grow even further to 2 000 to reach 36 Mtce. Some countries should become sizeable net importers (South Korea, Taiwan, and Philipines for coastal power plants) while ample exporting possibilities exist for Colombia, Venezuela, Mozambique, India and, to a lesser extent, Botswana.

The main obstacle to the greater realization of coal trade in LDCs lies in the timely development of infrastructure (particularly rail and road transport). The capital requirements will be large but probably can be financed given the potential for export or import substitution. The labour intensity of coal mining and handling operations should be an incentive to increase coal production and consumption in developing countries.

ANNEXES

Annex 1   Table 1-1
Coal Demand, Indigenous Supply and Trade, OECD, Reference Case[1]
Low Nuclear Assumption — Mtce

		Demand				Indi-genous Supply	Net Import (+)/Export (—)[1]			
		Metall-urgical [3]	Thermal Power St [2]	Coal Other	Total		Metall-urgical [4]	of which Coke	Thermal	Total
Canada	1976	7.8	14.4	1.7	23.9	20.1	—3.8	0.4	7.0	3.2
	1985	9.3	23.7	1.1	34.1	40.1	—14.6		8.6	—6.0
	1990	10.0	31.0	1.0	42.0	50.7	—17.3		8.6	—8.7
	2000	10.6	45.0	1.0	56.6	71.0	—23.0		8.6	—14.4
United States	1976	76.8	369.6	53.8	500.2	555.4	—43.2	—	—10.3	—53.5
	1985	87.9	615.6	65.8	769.3	837.3	—55.4		—12.6	—68.0
	1990	91.6	745.7	96.0	933.3	1 012.6	—60.7		—18.6	—79.3
	2000	91.6	800.0	160.8	1 052.4	1 181.0	—70.0		—58.6	—128.6
Japan	1976	68.5	7.3	7.4	83.2	20.4	57.1	—0.5	2.5	59.6
	1985	92.9	13.9	8.6	115.4	19.0	82.7		13.7	96.4
	1990	100.0	32.1	9.9	142.0	19.0	89.9		33.1	123.0
	2000	114.1	75.7	10.1	199.9	19.0	104.4		76.5	180.9
Australia	1976	9.3	24.8	3.2	37.3	69.6	—27.9	—0.2	—3.1	—31.0
	1985	12.1	37.6	2.2	51.9	108.9	—43.0		—14.0	—57.0
	1990	13.7	44.0	3.2	60.9	150.9	—54.0		—36.0	—90.0
	2000	17.3	69.7	3.1	90.1	285.1	—75.0		—120.0	—195.0
New Zealand	1976	—	0.7	1.7	2.4	2.4	—	—	—	—
	1985	—	0.4	1.9	2.3	2.3	—		—	—
	1990	—	1.4	2.3	3.7	3.7	—		—	—
	2000	—	4.3	3.4	7.7	7.7	—		—	—
Austria	1976	3.1	0.7	1.1	4.9	0.9	3.1	0.9	0.5	3.6
	1985	2.1	0.6	2.1	4.8	0.6	2.1		2.1	4.2
	1990	2.3	0.7	1.9	4.9	0.6	2.3		2.0	4.3
	2000	2.5	1.1	1.9	5.5	0.6	2.5		2.4	4.9
Belgium	1976	8.7	3.0	2.4	14.1	7.2	4.8	0.2	2.5	7.3
	1985	8.4	6.3	1.2	15.9	7.0	3.6		5.3	8.9
	1990	8.7	8.0	1.2	17.9	7.0	4.0		6.9	10.9
	2000	8.7	17.7	1.2	27.6	7.0	4.0		16.6	20.6
Denmark	1976	0.1	3.5	0.7	4.3	—	0.1	0.1	3.8	3.9
	1985	—	6.0	1.6	7.6	—	—		7.6	7.6
	1990	—	9.1	1.7	10.8	—	—		10.8	10.8
	2000	—	11.4	2.0	13.4	—	—		13.4	13.4
Finland	1976	0.9	2.9	0.3	4.1	—	0.9	0.9	2.6	3.5
	1985	0.9	2.7	0.5	4.1	—	0.9		3.2	4.1
	1990	1.0	2.7	1.0	4.7	—	1.0		3.7	4.7
	2000	1.1	3.4	2.2	6.7	—	1.1		5.6	6.7
France	1976	16.4	18.6	10.4	45.4	24.3	9.6	1.6	9.2	18.8
	1985	14.7	15.9	13.8	44.4	18.6	7.9		17.9	25.8
	1990	14.7	17.3	19.4	51.4	18.6	7.9		24.9	32.8
	2000	14.7	42.6	25.0	82.3	18.6	7.9		55.8	63.7
Germany	1976	35.0	67.1	10.0	112.1	126.1	—14.9	—5.5	3.2	—11.7
	1985	32.1	76.1	8.8	117.0	123.6	—13.6		7.0	—6.6
	1990	32.0	91.5	8.3	131.8	122.4	—12.9		22.3	9.4
	2000	38.3	138.7	13.3	190.3	125.1	—6.9		72.1	65.2
Greece	1976	0.5	4.0	0.6	5.1	4.4	0.5		0.1	0.6
	1985	1.3	9.7	0.6	11.6	10.3	1.3		—	1.3
	1990	1.7	11.9	0.8	14.4	12.6	1.8		—	1.8
	2000	2.2	17.7	1.0	20.9	18.7	2.2		—	2.2
Iceland	1976	—	—	—	—	—	—		—	—
	1985	—	—	—	—	—	—		—	—
	1990	—	—	—	—	—	—		—	—
	2000	—	—	—	—	—	—		—	—
Ireland	1976	—	—	0.6	0.6	—	—		0.6	0.6
	1985	—	—	0.9	0.9	—	—		0.9	0.9
	1990	—	—	0.9	0.9	—	—		0.9	0.9
	2000	—	4.0	0.9	4.9	—	—		4.9	4.9
Italy	1976	9.7	2.3	1.5	13.5	0.7	10.0	—0.8	1.5	11.5
	1985	10.1	7.1	3.3	20.5	2.1	10.1		8.3	18.4
	1990	11.9	11.5	3.5	26.9	2.3	11.9		12.7	24.6
	2000	14.0	19.7	4.5	38.2	2.6	14.0		21.6	35.6

1. In some instances net imports do not equal demand minus indigenous supply in 1976. Differences are due to stock changes and Statistical discrepencies.
2. Excludes metallurgical coal products used to produce electricity.
3. Domestic demand for metallurtical coal plus net imports (minus net exports) of coke.
4. Includes metallurgical coal and coke.

Annex 1   Table 1-1 (cont'd.)
Coal Demand, Indigenous Supply and Trade, OECD, Reference Case
Low Nuclear Assumption — Mtce

		Demand				Indi-genous	Net Import (+)/Export (—)[1]			
		Metall-urgical [3]	Thermal Power St[2]	Coal Other	Total	Supply	Metall-urgical [4]	*of which* Coke	Thermal	Total
Luxembourg	1976	2.0	—	0.7	2.7	—	2.0	2.0	0.6	2.6
	1985	3.1	—	1.0	4.1	—	3.1		1.0	4.1
	1990	3.3	—	1.1	4.4	—	3.3		1.1	4.4
	2000	3.7	—	1.2	4.9	—	3.7		1.2	4.9
Netherlands	1976	3.1	0.9	0.3	4.3	—	3.2	—0.4	1.1	4.3
	1985	3.6	4.1	0.7	8.4	—	3.6		4.8	8.4
	1990	4.3	11.8	2.4	18.5	—	4.3		14.2	18.5
	2000	4.7	30.0	3.7	38.4	—	4.7		33.7	38.4
Norway	1976	0.8	—	0.5	1.3	0.5	0.9	0.4	—0.1	0.8
	1985	0.6	—	1.1	1.7	1.4	0.6		—0.3	0.3
	1990	0.7	—	1.3	2.0	1.4	0.7		—0.1	0.6
	2000	0.9	1.0	1.4	3.3	1.3	0.9		1.1	2.0
Portugal	1976	0.3	0.2	0.1	0.6	0.2	0.4		—	0.4
	1985	0.6	1.3	—	1.9	0.3	0.6		1.0	1.6
	1990	0.7	2.2	—	2.9	0.3	0.7		1.9	2.6
	2000	1.0	5.0	—	6.0	0.3	1.0		4.7	5.7
Spain	1976	6.7	6.1	1.9	14.7	10.6	4.7	0.3	0.2	4.9
	1985	8.4	12.1	2.2	22.7	18.6	4.1		—	4.1
	1990	9.1	15.3	2.2	26.6	22.0	4.6		—	4.6
	2000	9.4	23.7	2.3	35.4	25.0	4.8		5.6	10.4
Sweden	1976	2.5	—	—	2.5	—	2.6	1.0	0.4	3.0
	1985	2.0	—	2.4	4.4	—	2.0		2.4	4.4
	1990	2.4	—	2.6	5.0	—	2.4		2.6	5.0
	2000	2.9	1.6	2.8	7.3	—	2.9		4.4	7.3
Switzerland	1976	—	—	0.3	0.3	—	—	—	0.3	0.2
	1985	—	0.1	0.5	0.6	—	—		0.6	0.6
	1990	—	0.4	0.5	0.9	—	—		0.9	0.9
	2000	—	0.9	0.6	1.5	—	—		1.5	1.5
Turkey	1976	2.1	2.1	2.6	6.8	6.8	—		—	—
	1985	6.9	11.4	16.6	34.9	27.9	4.4		2.6	7.0
	1990	11.4	16.5	22.1	50.0	33.9	9.0		7.1	16.1
	2000	16.1	26.5	25.1	67.7	44.4	13.7		9.6	23.3
United Kingdom	1976	20.3	62.8	18.4	101.5	103.0	0.3	—1.0	0.4	0.7
	1985	22.0	70.7	16.9	109.6	111.0	—		—1.4	—1.4
	1990	22.5	71.7	16.9	111.1	111.1	—		—	—
	2000	24.0	74.0	22.4	120.4	120.4	—		—	—
OECD Europe	1976	112.2	174.2	52.4	338.8	284.7	28.2	—0.3	26.9	55.1
	1985	116.8	224.1	74.2	415.1	321.4	30.7		63.0	93.7
	1990	126.7	270.6	87.8	485.1	332.2	41.0		111.9	152.9
	2000	144.2	419.0	111.5	674.7	364.0	56.5		254.2	310.7
OECD Total	1976	274.6	591.0	120.2	985.8	952.6	10.4	—0.6	23.0	33.4
	1985	319.0	915.3	153.8	1 388.1	1 329.0	0.4		58.7	59.1
	1990	342.0	1 124.8	200.2	1 667.0	1 569.1	—1.1		99.0	97.9
	2000	377.8	1 413.7	289.9	2 081.4	1 927.8	—7.1		160.7	153.6
IEA Total	1976	247.7	544.5	106.2	898.4	858.5	27.4	—2.9	14.2	41.6
	1985	290.7	857.8	137.3	1 285.8	1 201.2	34.0		50.6	84.6
	1990	311.9	1 058.6	176.6	1 547.1	1 399.3	43.3		104.5	147.8
	2000	343.7	1 293.0	259.6	1 896.3	1 623.8	57.9		214.6	272.5

Annex 1 Table 1-2
Electricity Production and Energy Used in Electricity Generation, OECD, Reference Case
Mtce

Country	Year	Elect. Prod'n	Low Nuclear — Coal¹	Oil	Gas	Nuclear	Hydro/Geo	Other	High Nuclear — Coal	Oil	Gas	Nuclear	Hydro/Geo	Other
Canada	1976	37.0	14.4	4.7	6.3	6.6	80.3	—						
	1985	53.9	23.7	9.6	7.6	24.1	101.0	—	23.7	9.6	7.6	24.1	101.0	—
	1990	64.6	31.0	8.9	8.0	34.7	114.3	—	29.1	8.9	8.0	38.6	112.3	—
	2000	90.1	45.0	8.9	8.0	85.5	125.4	—	39.4	8.9	8.0	94.1	122.4	—
United States	1976	285.3	369.6	120.2	104.4	66.2	95.8	—						
	1985	421.3	615.6	160.3	57.1	208.6	117.7	4.3	615.6	140.9	57.1	228.0	117.7	4.3
	1990	497.4	745.7	146.9	54.3	292.0	139.4	14.3	702.1	125.0	54.3	357.5	139.4	14.3
	2000	604.9	800.0	100.0	32.9	547.3	157.1	42.9	725.7	70.0	32.9	651.6	157.1	42.9
Japan	1976	62.9	17.3	100.5	7.3	11.9	30.9	—						
	1985	105.3	29.4	161.7	47.1	35.1	39.1	—	29.4	157.8	47.1	39.0	39.1	—
	1990	132.7	48.8	163.4	63.3	65.1	46.9	—	48.8	153.9	63.3	74.6	46.9	—
	2000	167.4	94.7	108.1	66.7	147.1	55.4	—	64.3	101.7	66.7	183.9	55.4	—
Australia	1976	9.4	24.8	0.8	1.2	—	6.8	—						
	1985	14.4	37.6	1.1	3.0	—	6.4	—	37.6	1.1	3.0	—	6.4	—
	1990	18.3	44.0	1.0	3.0	5.0	8.1	—	41.5	1.0	3.0	7.5	8.1	—
	2000	27.3	69.7	1.1	3.0	10.0	8.6	—	64.7	1.1	3.0	15.0	8.6	—
New Zealand	1976	2.6	0.8	0.5	0.7	—	5.4	—						
	1985	3.7	0.4	0.1	2.4	—	7.9	—	0.4	0.1	2.4	—	7.9	—
	1990	4.6	1.4	0.1	3.3	—	8.3	—	1.4	0.1	3.3	—	8.3	—
	2000	6.3	4.3	0.1	3.3	2.1	8.3	—	4.3	0.1	3.3	2.1	8.3	—
Austria	1976	4.3	0.8	1.7	1.3	—	6.7	—						
	1985	6.4	0.6	2.1	2.3	1.3	10.4	—	0.6	2.1	2.3	1.3	10.4	—
	1990	8.0	0.7	4.9	2.7	1.3	11.1	—	0.7	4.9	2.7	1.3	11.1	—
	2000	11.7	1.1	3.7	4.3	8.1	13.0	—	1.1	3.7	2.5	11.8	11.1	—
Belgium	1976	5.8	3.9	5.0	2.6	3.1	0.1	—						
	1985	9.3	6.3	5.1	2.9	9.0	0.4	—	6.3	5.1	2.9	9.0	0.4	—
	1990	11.9	8.0	7.6	1.4	11.4	0.4	—	8.0	4.6	1.4	14.4	0.4	—
	2000	18.7	17.7	8.6	—	18.7	0.4	—	8.5	8.6	—	27.9	0.4	—
Denmark	1976	2.6	3.5	3.6	—	—	—	0.8						
	1985	3.3	6.0	2.6	—	—	—	1.0	6.0	2.6	—	—	—	—
	1990	3.9	9.1	1.0	—	—	—	1.0	4.3	1.0	—	4.9	—	—
	2000	4.6	11.4	0.4	—	—	—	1.0	3.7	0.4	—	7.7	—	—
Finland	1976	3.6	2.9	2.5	0.4	—	3.3	0.8						
	1985	5.3	2.7	1.1	0.3	4.4	5.3	1.0	2.7	1.1	0.3	4.4	5.3	1.0
	1990	6.7	2.7	2.4	0.3	7.0	5.3	1.0	2.7	2.4	0.3	7.0	5.3	1.0
	2000	10.1	3.4	3.9	0.3	13.9	5.3	1.0	2.4	1.0	0.3	17.8	5.3	1.0

Annex 1 Table 1-2 (cont'd.)

Electricity Production and Energy Used in Electricity Generation, OECD, Reference Case

Mtce

		Low Nuclear							High Nuclear					
		Elect. Prod'n	Coal[1]	Oil	Gas	Nuclear	Hydro/Geo	Other	Coal	Oil	Gas	Nuclear	Hydro/Geo	Other
France	1976	25.0	20.8	19.5	2.9	4.9	15.5	—						
	1985	42.3	15.9	10.4	1.6	60.4	20.6	4.3	15.9	10.4	1.6	60.4	20.6	4.3
	1990	56.6	17.3	9.6	1.6	90.9	20.4	6.3	17.3	9.6	1.6	90.9	20.4	6.3
	2000	100.7	42.6	7.9	1.6	166.9	20.4	15.7	23.7	7.9	1.6	185.8	20.4	15.7
Germany	1976	41.0	67.1	10.6	17.4	7.9	4.7	3.8						
	1985	63.9	76.1	21.0	24.3	30.9	7.0	2.4	75.9	17.6	23.3	37.8	7.0	2.4
	1990	80.9	91.5	20.1	24.3	52.9	7.0	5.0	—	14.1	22.1	71.9	7.0	5.0
	2000	111.1	138.7	15.0	20.0	101.9	8.0	11.0	94.2	5.0	10.0	169.3	8.0	11.0
Greece	1976	2.2	4.0	2.0	—	—	0.7	—						
	1985	4.4	9.7	2.3	—	—	1.7	—	9.7	2.3	—	—	1.7	—
	1990	6.6	11.9	5.1	—	1.4	1.4	—	11.9	5.1	—	1.4	1.4	—
	2000	13.0	17.7	12.7	—	7.7	1.4	—	17.7	5.0	—	15.4	1.4	—
Iceland	1976	0.3	—	—	—	—	0.8	—						
	1985	0.4	—	0.1	—	—	1.0	—	—	0.1	—	—	1.0	—
	1990	0.6	—	0.1	—	—	1.3	—	—	0.1	—	—	1.3	—
	2000	0.9	—	0.1	—	—	2.3	—	—	0.1	—	—	2.3	—
Ireland	1976	1.1	—	1.9	0.9	—	0.3	0.9						
	1985	2.0	—	3.7	0.9	—	0.3	1.0	—	3.7	0.9	—	0.3	—
	1990	3.0	—	6.9	0.9	—	0.3	1.0	—	6.9	0.9	—	0.3	—
	2000	5.1	4.0	6.3	—	2.1	0.3	1.0	1.9	6.3	0.9	4.2	0.3	4.1
Italy	1976	20.1	3.4	28.4	4.6	1.2	13.5	—						
	1985	34.4	8.3	54.1	3.0	5.6	17.1	—	8.3	50.7	3.0	9.0	17.1	—
	1990	41.3	12.9	69.3	0.4	9.0	17.1	—	12.9	56.0	0.4	22.3	17.1	—
	2000	64.4	21.4	64.6	1.1	60.9	17.4	4.1	16.4	43.7	1.1	86.8	17.4	4.1
Luxembourg	1976	0.1	0.2	0.1	0.2	—	0.2	—						
	1985	0.9	0.4	1.4	0.1	—	0.4	—	0.4	1.4	0.1	—	0.4	—
	1990	1.4	0.4	3.1	0.1	—	0.4	—	0.4	3.1	0.1	—	0.4	—
	2000	2.4	0.9	1.9	0.1	3.5	0.7	—	0.9	0.1	0.1	5.3	0.7	—
Netherlands	1976	7.1	1.4	1.1	13.8	1.2	—	—						
	1985	8.7	4.7	9.6	8.1	1.3	—	—	4.7	9.6	8.1	1.3	—	—
	1990	11.1	12.5	12.9	4.3	1.3	—	—	12.5	12.9	4.3	1.3	—	—
	2000	16.6	30.8	14.0	—	1.3	—	0.4	23.4	8.0	—	14.7	—	0.4
Norway	1976	10.1	—	—	—	—	17.6	—						
	1985	11.4	—	—	—	—	19.9	—	—	—	—	—	19.9	—
	1990	12.4	—	—	—	—	21.9	—	—	—	—	—	21.9	—
	2000	14.7	1.0	1.0	—	—	23.7	—	2.0	2.0	—	—	21.7	—

Annex 1 Table 1-2 (cont'd.)

Electricity Production and Energy Used in Electricity Generation, OECD, Reference Case

Mtce

		Low Nuclear							High Nuclear					
		Elect. Prod'n	Coal[1]	Oil	Gas	Nuclear	Hydro/Geo	Other	Coal	Oil	Gas	Nuclear	Hydro/Geo	Other
Portugal	1976	1.2	0.2	1.5	—	—	1.6	—						
	1985	2.7	1.3	1.7	—	—	4.3	—	1.3	1.7	—	—	4.3	—
	1990	4.0	2.1	3.9	—	—	4.9	—	2.1	3.1	—	0.8	4.9	—
	2000	7.7	5.0	3.9	—	5.3	6.3	—	3.0	2.6	—	8.6	6.3	—
Spain	1976	11.2	6.2	15.1	0.5	2.5	7.4	—						
	1985	16.6	12.1	13.1	—	10.7	13.1	—	12.1	13.1	—	10.7	13.1	—
	1990	21.0	15.3	12.9	—	20.3	13.6	—	15.3	6.8	—	26.4	13.6	—
	2000	32.0	23.7	5.0	—	50.7	14.7	—	15.3	2.0	—	62.8	14.7	—
Sweden	1976	10.7	—	3.6	—	5.3	18.0	0.1						
	1985	14.7	—	8.9	—	7.3	20.6	0.4	—	2.9	—	13.3	20.6	0.4
	1990	17.0	—	6.0	—	14.1	21.9	0.6	—	2.2	—	18.8	21.0	0.6
	2000	21.6	1.6	6.0	—	22.1	23.7	0.9	—	2.0	—	27.7	23.7	0.9
Switzerland	1976	4.6	—	0.6	—	2.6	8.8	—						
	1985	5.9	0.1	3.4	0.1	2.6	10.0	—	0.1	1.7	0.1	4.3	10.0	—
	1990	6.7	0.4	1.5	0.1	5.6	10.3	—	—	1.1	0.1	6.7	10.0	—
	2000	8.3	0.9	1.5	0.1	10.4	9.9	0.1	—	0.4	0.1	11.9	9.9	0.1
Turkey	1976	2.2	2.3	2.3	—	—	2.7	—						
	1985	6.9	12.1	1.9	—	—	4.3	—	12.1	1.3	—	0.6	4.3	—
	1990	10.0	17.6	1.6	—	0.6	6.9	—	17.6	1.6	—	0.6	6.9	—
	2000	18.4	28.1	3.6	—	6.0	11.1	0.1	24.1	3.6	—	10.0	11.1	0.1
United Kingdom	1976	34.0	63.4	17.1	2.6	12.7	1.8	—						
	1985	40.7	71.4	14.4	2.9	19.7	1.4	—	71.4	14.4	2.9	19.7	1.4	—
	1990	44.4	72.4	14.3	2.9	25.7	1.4	—	72.4	12.1	2.9	27.9	1.4	—
	2000	57.1	74.7	24.3	—	50.0	1.4	—	74.7	14.3	—	60.0	1.4	—
OECD Europe	1976	187.2	180.1	116.6	46.3	41.4	103.7	5.6	180.1	116.6	46.3	41.4	103.7	8.6
	1985	280.2	227.7	156.9	46.5	153.2	137.8	9.1	227.5	141.8	45.5	171.8	137.8	9.1
	1990	347.5	274.8	183.2	39.0	241.5	145.6	13.9	267.2	147.6	36.8	296.6	144.4	13.9
	2000	519.1	424.7	183.9	28.4	529.5	160.0	34.2	313.0	116.7	16.6	727.7	156.1	34.2
OECD Total	1976	584.4	607.0	343.3	166.2	126.1	322.9	5.6	607.0	343.3	166.2	126.1	322.9	5.6
	1985	878.8	934.4	489.7	163.7	421.0	409.9	13.4	934.2	451.3	162.7	462.9	409.9	13.4
	1990	1 065.1	1 145.7	503.5	170.9	638.3	462.6	28.2	1 090.1	436.5	168.7	774.8	459.4	28.2
	2000	1 415.1	1 438.4	402.1	142.3	1 321.5	514.8	77.1	1 211.4	298.5	130.5	1 674.4	507.9	77.1
IEA Total	1976	544.9	558.3	319.0	161.7	121.2	294.9	4.8	558.3	319.0	161.7	121.2	294.9	4.8
	1985	813.7	876.9	475.3	158.8	356.2	372.3	8.1	876.7	436.9	157.8	398.1	372.3	8.1
	1990	978.9	1 079.6	486.5	166.0	535.4	422.6	20.9	1 026.5	420.3	163.8	668.6	419.4	20.9
	2000	1 268.4	1 317.5	385.2	137.4	1 125.4	471.9	60.4	1 117.6	285.8	125.6	1 447.2	465.0	60.4

1. Includes metallurgical coal products used to produce electricity.

Annex 1   Table 1-3
World Coal Trade

Mtce
Reference Case
Net Imports (+)/Exports (—)

	1976		1985		1990		2000	
North America	—50.3		—74		—88		—143	
Thermal		—3.3		—4		—10		—50
Coking		—47.0		—70		—78		—93
OECD Europe	55.1		94		153		311	
Thermal		26.9		63		112		254
Coking		28.2		31		41		57
Japan	59.6		96		123		181	
Thermal		2.5		14		33		77
Coking		57.1		83		90		104
Australia	—31.0		—57		—90		—195	
Thermal		—3.1		—14		—36		—120
Coking		—27.9		—43		—54		—75
OECD Total	33.4		59		98		154	
Thermal		23.0		59		99		161
Coking		10.4		—		—1		—7
CPE's[1]	—37.7		—43		—49		—66	
Thermal		—18.3		—23		—26		—41
Coking		—19.4		—20		—23		—25
LDC's[2]	7		11		2		—10	
Thermal		—		—5		—17		—36
Coking		7		16		—19		26
South Africa	—5.6		—34		—60		—90	
Thermal		—4.8		—34		—60		—90
Coking		—0.8		—		—		—
Other[3]	2.5		7		9		12	
Thermal		—0.1		3		4		6
Coking		2.6		4		5		6

Partial sums may not add up to totals due to rounding errors.
1. Centrally Planned Economies of Eastern Europe and Asia (does not include Yugoslavia).
2. Less Developed Countries except Turkey, included in OECD Europe.
3. Including Yugoslavia, Israel and other smaller countries not comprised in previous groupings.

Annex 1 Table 1-4

Fuel Inputs for Electricity Generation in the Reference and Enlarged Coal Cases

Mtce

	Elect. Prod'n	Low Nuclear						High Nuclear					
		Coal	Oil	Gas	Nuclear	Hydro-Geo	Other	Coal	Oil	Gas	Nuclear	Hydro-Geo	Other
North America													
1976	322.3	384.0	124.9	110.7	72.8	176.1	—						
1985 reference	475.2	639.3	169.9	64.7	232.7	218.7	4.3	639.3	150.5	64.7	252.1	218.7	4.3
1985 enlarged	475.2	665.8	150.5	57.6	232.7	218.7	4.3	646.4	150.5	57.6	252.1	218.7	4.3
1985 change (+ or —)		+26.5	−19.4	−7.1				+7.1		−7.1			
1990 reference	562.0	776.7	155.8	62.3	326.7	253.7	14.3	731.2	133.9	62.3	396.1	251.7	14.3
1990 enlarged	562.0	848.6	108.9	37.3	326.7	253.7	14.3	781.2	108.9	37.3	396.1	251.7	14.3
1990 change (+ or —)		+71.9	−46.9	−25.0				+50.0	−25.0	−25.0			
2000 reference	695.0	845.0	108.9	40.9	632.8	282.5	42.9	765.1	78.9	40.9	745.7	279.5	42.9
2000 enlarged	695.0	912.9	58.9	23.0	632.8	282.5	42.9	803.0	58.9	23.0	745.7	279.5	42.9
2000 change (+ or —)		+67.9	−50	−17.9				+37.9	−20	−17.9			
OECD Pacific													
1976	74.9	42.9	101.8	9.2	11.9	43.1							
1985 reference	123.4	67.4	162.9	52.5	35.1	53.4		67.4	159.0	52.5	39.0	53.4	
1985 enlarged	123.4	71.3	159.0	52.5	35.1	53.4		67.4	159.0	52.5	39.0	53.4	
1985 change (+ or —)		+3.9	−3.9										
1990 reference	155.6	94.2	164.5	69.6	70.1	63.3		91.7	155.0	69.9	82.1	63.3	
1990 enlarged	155.6	107.6	151.1	69.6	70.1	63.3		100.6	146.1	69.9	82.1	63.3	
1990 change (+ or —)		+13.4	−13.4					+8.9	−8.9				
2000 reference	201.0	168.7	109.3	73.0	159.2	72.3		133.3	102.9	73.0	201.0	72.3	
2000 enlarged	201.0	196.8	81.2	73.0	159.2	72.3		166.7	76.2	66.3	201.0	72.3	
2000 change (+ or —)		+28.1	−28.1					+33.4	−26.7	−6.7			
OECD Europe													
1976	187.3	180.1	116.6	46.3	41.4	103.7	5.6						
1985 reference	280.2	227.7	156.9	46.5	153.2	137.8	9.1	227.5	141.8	45.5	171.8	137.8	9.1
1985 enlarged	280.2	244.6	140.0	46.5	153.2	137.8	9.1	239.3	130.0	45.5	171.8	137.8	9.1
1985 change (+ or —)		+16.9	−16.9					+11.8	−11.8				

Annex 1 Table 1-4 (cont'd.)

Fuel Inputs for Electricity Generation in the Reference and Enlarged Coal Cases

Mtce

	Elect. Prod'n	Low Nuclear						High Nuclear					
		Coal	Oil	Gas	Nuclear	Hydro-Geo	Other	Coal	Oil	Gas	Nuclear	Hydro-Geo	Other
1990 reference	347.5	274.8	183.2	39.0	241.5	145.6	13.9	267.2	147.6	36.8	296.6	144.4	13.9
enlarged	347.5	298.0	160.0	39.0	241.5	145.6	13.9	284.8	130.0	36.8	296.6	144.4	13.9
change (+ or —)		+23.2	−23.2					+17.6	−17.6				
2000 reference	519.1	424.7	183.9	28.4	529.5	160.0	34.2	313.0	116.7	16.6	727.7	156.1	34.2
enlarged	519.1	508.6	100.0	28.4	529.5	160.0	34.2	354.7	75.0	16.6	727.4	156.1	34.2
change (+ or —)		+83.9	−83.9					+41.7	−41.7				
OECD Total													
1976	584.5	607.0	343.3	166.2	126.1	322.9	5.6						
1985 reference	878.8	934.4	489.7	163.7	421.0	409.9	13.4	934.2	451.3	162.7	462.9	409.9	13.4
enlarged	878.8	981.7	449.5	156.6	421.0	409.9	13.4	953.1	439.5	155.6	462.9	409.9	13.4
change (+ or —)		+47.3	−40,2	−7.1				+18.9	−11.8	−7.1			
1990 reference	1 065.1	1 145.7	503.5	170.9	638.3	462.6	28.2	1 090.1	436.5	169.0	774.8	459.4	28.2
enlarged	1 065.1	1 254.2	420.0	145.9	638.3	462.6	28.2	1 166.6	385.0	144.0	774.8	459.4	28.2
change (+ or —)		+108.5	−83.5	−25.0				+76.5	−51.5	−25.0			
2000 reference	1 415.1	1 438.4	402.1	142.3	1 321.5	514.8	77.1	1 211.4	298.5	130.5	1 674.5	507.3	77.1
enlarged	1 415.1	1 618.3	240.1	124.4	1 321.5	514.8	77.1	1 324.4	210.1	105.9	1 674.4	507.3	77.1
change (+ or —)		+179.9	−162.0	−17.9				+113.0	−88.4	−24.6			
Memorandum Item :													
OECD Total in Mtoe													
2000 reference	990.6	1 006.9	281.5	99.6	925.1	360.4	54.0	845.0	209.0	91.4	1 172.2	355.1	54.0
enlarged	990.6	1 132.8	168.1	87.1	925.1	360.4	54.0	927.1	147.1	74.1	1 172.2	355.1	54.0
change (+ or —)		+125.9	−113.4	−12.5				+79.1	−61.9	−17.2			

Annex 2

TRENDS IN OECD STEAM COAL TRADE

International Hard Coal Trade

International trade in hard coal has grown to reach some 195 Mt in 1975 or some 8 per cent of total world output, from some 100 Mt (or 5 per cent of world output) in 1960. A major portion of the world trade in hard coal is carried out within the EEC, the COMECON or between the U.S. and Canada. Annex 2, Table 2-1 shows hard coal exports in selected countries in the 1960-1977 period.

OECD Thermal Coal Trade

An attempt to analyse the effects of the oil price rise on international coal trade is a difficult task since there are no international statistics to show a breakdown of coal exports/imports by type of use. Therefore, it is impossible to present official figures on coking coal or thermal coal trade. However, some estimates can be produced using various sources. Annex 2, Table 2-2, shows an estimated breakdown of OECD hard coal imports, exluding intra EEC trade, by types of coal — coking and thermal. OECD thermal coal trade appears to have reached a highpoint in the mid-sixties, decreasing to a low of 27 Mt in 1973. From 1973 on, OECD thermal coal imports have increased continuously through 1976 in which year they reached 40 Mt.

When only steam coal imports are considered (that is thermal coal used in power generation) the rapid growth in imports following the 1973-74 oil price rise is evident. In the OECD European region imports of steam coal for power generation (excluding intra EEC trade) were in 1977 2.5 times greater than in 1973.

A more detailed breakdown of OECD thermal coal imports in the 1973-76 period is shown in Annex 2, Table 2-3.

The analysis of data shown in Tables 2-2 and 2-3 presents several interesting points. First, thermal coal imports, which represented over 60 per cent of total hard coal imports into OECD countries in 1960, started by the mid-sixties to lose relative importance. By 1973 the only OECD thermal coal trade which remained was:

a) U.S. coal exports to Canada of both anthracite and bituminous coal. Almost all this coal was used by Ontario Hydro for the production of electric power. The volumes traded are fairly uniform.

b) Some 1.0 Mt of anthracite imports to Japan mostly from Korea, North Vietnam and China. Steam coal imports did not start in Japan until 1974 due to trade policies, which did not permit them.

c) The European market for imported thermal coal diminished to just under 18 Mt supplied mostly by Poland (some 10 Mt) and the Soviet Union (some 3 Mt). These imports were of both anthracite (delivered mainly to the domestic/residential sector) and bituminous coal. Almost 10 Mt were used for electric power generation. The U.S. was also an important supplier to Europe, particularly to Germany. It is worth noting that most of the coal exported to Germany from the U.S. was of metallurgical quality although it was not utilized in coke ovens. This coal could have been utilized in the few remaining gas works or by the industrial sector which converted it into coke. Data available does not permit verification of such uses. Outside the EEC, Finland was the most important thermal coal importer.

In 1974 following the oil price rise, thermal coal imports rose to 24 Mt in OECD Europe. Most of this increase occurred in the electricity generation sector of EEC-9. Conditions favoured such a shift due to the existence of dual-fired electric generating capacity, fueled mainly by fuel-oil until 1973 and to some surplus coal production capacity worldwide. Poland provided the major portion of the additional coal supplies, but Australia and South Africa began shipments to the European market. The British Central Electricity Generating Board received the bulk of the Australian thermal coal exports. British imports rose as a consequence of domestic supply disruption due to strikes in the mining industry.

Germany increased her imports of thermal coal up to 4.7 Mt in 1974. Unfortunately, it is not known how much of this coal was used for power generation. Japan started to import steam coal from Australia, China and the USSR for the first time after the Government lifted existing bans on steam coal imports. In Japan, a token quantity of 115 000 t was used for electricity generation as tests were carried out to study the possibility of expanded steam coal imports.

In 1975, European imports of thermal coal increased to 28 Mt. The additional imports were consumed in power stations. British imports in 1975 also increased for the same reasons described above. In Germany public utilities consumed 3.8 Mt as coal stocks at collieries were rapidly rising. More significant was the increase in French imports which reached almost 7.0 Mt in 1975 compared to 4.5 Mt in 1974, all the increase going to the power generation sector. In the other OECD regions, the coal situation remained stable with the exception of Canada where imports increased after a slump in 1974. However, this was due to stocking and de-stocking operations more than to actual use of imported coal by Ontario Hydro. Japan continued tests on the use of imported coal for power generation.

The main source for the EEC increase in thermal coal imports was the U.S. and Canada as Poland and the USSR maintained their deliveries almost constant. South Africa continued her penetration into the world markets as Richards Bay Coal Terminal became operational, despite some internal opposition to exporting coal.

1976 witnessed a further increase in thermal coal imports, again marked by increased deliveries to power generating stations. The British Central Electricity Generating Board reduced its imports from Australia to 1.3 Mt. German imports decreased slightly. The largest increase in imports was shown by the French Public Utility, EDF, whose consumption of imported steam coal rose to almost 7 Mt, more than one third of all coal used in France for electricity generation. Denmark imported 4.2 Mt, about the same as the year before. However, it should be pointed out that Denmark was probably using as much coal as possible in power generation. For the first time the Netherlands burned sizeable quantities of imported coal (0.8 Mt) in power stations. Polish coal deliveries to theEEC increased again to 11.3 Mt as well as imports from the USSR. A large increase in South African exports brought this country into the limelight as one of the most probable coal suppliers in the mid-term. Imports from the U.S. decreased probably as a consequence of conditions in the spot freight market. Thermal coal imports from Australia were reduced by half from the previous year.

In 1977 coal imports into the EEC from third-party countries increased 0.8 Mt to 44.5 Mt, of which only 18.1 Mt were delivered to cokeries (down 1.9 Mt from the 1976 level). Thermal coal imports, thus, increased by 2.6 Mt to 26.4 Mt, an 11 per cent growth (9.9 Mt for France, up 43 per cent from 1976; 4.2 Mt for Germany, up 17 per cent, and 3.6 Mt for Denmark, up 8 per cent, to mention the most important imported steam coal users). South Africa provided 7.6 Mt; Poland 10.4 Mt, the USSR

2.9 Mt and Australia and the U.S.A. 2.5 Mt each. Deliveries to public power stations increased by 3.8 Mt to 21.6 Mt. Taking into account steam coal imports from third-party countries for industrial electricity production and other OECD European imports, steam coal imports into OECD Europe could have reached 26 Mt.

Short Term Outlook

The short term outlook for coal consumption and international trade are determined by the prospects for iron and steel demand and power plants under construction. The trend toward smaller direct use of coal by industrial and residential energy consumers should continue, although the rate of decrease might be somehow reduced (the projected mandatory conversion of large industrial boilers in the U.S. may be an important exception to the general decline in direct coal consumption). Throughout the OECD prospects for coking coal consumption in the short term are rather poor. Coal consumption in coke ovens and coking coal trade are expected to remain at the same level of 1976 through the early eighties. Contribution to increases in coal consumption and international trade in the short term should be concentrated in the power generating sector in the OECD region and elsewhere in the world outside communist areas. (The only probable exception would be cement factories whose consumption of imported coal may increase in the short term).

Increased coal consumption by the electric power sector has been shown to be the reason for increased coal imports into the EEC since 1973 by intensive use of dual-fired capacity previously fueled by oil products. By the end of 1974 the EEC had some 99.9 GW net output capacity which could be fired by solid fuels out of a total of 165.9 GW net capacity. Roughly a third of that mentioned solid fueled capacity was in combined (bivalent or trivalent) power stations which were using liquid or gaseous hydrocarbons by 1974. Of all the capacity commissioned during 1974 in EEC member countries, capacity to burn coal along with other fuels outside Germany and the UK was limited to some 550 MW in Denmark. The same situation continued in 1975 as new stations were commissioned which had been started prior to 1973. The increase in coal use in power generation has been possible by increased use of old coal-fired capacity in the system base load. Many of the stations so used are more than twenty years old and despite a likely delay in decommissioning, a fair amount of coal-fueled capacity built in Europe in the reconstruction period which followed the end of the war will not be available in the short term. During 1976, for example, some 10 000 MW were decommissioned.

New coal-fired plants on stream could not upset this decrease so that by the end of 1976 only 94.3 GW could in principle be fired by coal in the EEC area. According to stated plans at the end of 1976 for the 1977-82 period some 38.2 GW thermal conventional capacity would come on stream in EEC countries, of which only 6.8 GW will have coal burning capabilities. Furthermore only 2.1 GW will be commissioned in countries likely to import steam coal. There is however a number of power stations planned to burn oil products where introduction of dual-fired capacity could be envisaged before 1982. This is particularly true of a fair amount of the 31 GW marked to burn liquid or gaseous hydrocarbons to start generating in the 1977-82 period.

Elsewhere in OECD Europe the situation is similar with no planned power capacity to burn imported coal although policy changes could be expected particularly in Scandinavian countries.

In conclusion if no policy changes are implemented in the near future to mandate conversion of new planned capacity in OECD Europe from oil to oil and coal wherever possible, and to retrofit other power stations (providing auxiliary equipment and/or infrastructure in existing power stations designed to burn coal but unable nowadays to do so), the present levels of steam coal imports into OECD Europe will hardly be maintained through the early 80s.

In Japan the very small increase in steam coal imports has been led by the industrial users (mainly cement factories). However, some planned power capacity has already been committed to imported coal. As of 1977 it was forecast that by 1985 some 5.5 GW of new coal-fired capacity will be fueled with imported steam coal. That should gradually increase thermal coal imports to Japan from the present level of some 2.0 Mt per year to some 5.0 Mt in 1980 and 15-16 Mt by 1985.

Elsewhere in the world plans have been announced to build new imported-coal fired power stations in Israel (1 400 MW by 1989) using South African coal. Other developing countries in East Asia were encouraging the construction of new power plants committed to imported coal.

On the supply side developments in the international steam coal trade will in the short term be dominated by the increase in South African steam coal exports which, in the absence of any substantial upward trend in world steam coal trade, could take over markets in Europe nowadays occupied by North American coal.

Annex 2   Table 2-1   **Hard Coal Exports in Selected Countries**
Mt

	1960	1970	1973	1974	1975	1976	1977*
USA	34.5	65.1	49.0	55.4	60.6	54.5	49.3
Canada	0.8	4.0	10.9	10.7	11.1	11.8	12.0
UK	5.3	3.2	2.7	1.6	1.9	1.4	1.8
Germany	17.8	15.8	14.0	17.0	14.7	13.0	13.7
Australia[1]	1.6	18.3	28.1	27.8	29.9	34.1	33.6
USSR	12.3	24.5	24.5	26.2	26.1	26.8	28.2
Poland	17.5	28.8	35.9	40.1	38.5	38.9	39.3
South Africa	1.0	1.5	1.9	2.3	2.7	6.0	12.7

Source:
OECD Basic Energy Statistics.
UN ECE: Coal Statistics.
South Africa: "Department of Energy and Environment."
International coal trade.

1. Fiscal year.
* Preliminary figures.

Annex 2   Table 2-2
Hard Coal Imports into OECD Regions and Estimates of Steam Coal Imports
Mt

	1960	1969	1973	1974	1975	1976	1977*
OECD Total							
Hard Coal	51.8	95.3	114.3	130.4	133.2	133.2	
Thermal Coal	33.0	33.0	26.8	33.1	39.4	39.7	
Supplies to Power Stations	n.a.	n.a.	17.2	19.1	29.8	30.8	
North America							
Hard Coal	12.5	14.8	15.0	14.3	16.3	14.6	
Thermal Coal	8.3	8.8	8.0	7.5	9.8	9.3	
Supplies to Power Stations	n.a.	n.a.	7.5	7.2	9.4	8.9	
OECD Europe[1]							
Hard Coal	30.6	37.2	41.3	51.5	54.6	57.4	
Thermal Coal	23.5	20.9	17.8	24.0	28.0	27.6	
Supplies to Power Stations	n.a.	n.a.	9.8	13.7	20.7	21.7	26.0
Pacific (Japan)							
Hard Coal	8.7	43.3	58.0	64.6	62.3	60.1	60.8
Thermal Coal	1.5	3.3	2.0	2.8	2.5	2.4	3.0
Supplies to Power Stations	—	—	—	0.1	0.2	0.2	0.3

1. Excluding intra-community trade (EEC).
* Preliminary data.

Annex 2 Table 2-3　Thermal[1] Coal Imports of OECD Countries (1973-76)

Mt

a) 1973

Importing Area	Exporting Area							Total thermal coal imports	Delivered to electric sector	Total hard coal imports
	USA	Canada	Australia	South Africa	USSR	Poland	Other			
EEC[2]	3.0		0.4	0.7	2.2	6.7	0.8	13.8	8.0	29.8
of which:										
Denmark	—	—	—	—	0.4	2.6	—	3.0	2.4	3.0
France	0.1	—	—	—	1.1	1.1	—	2.8	1.2	5.4
Germany	1.9	—	0.1	0.5	0.1	1.8	0.5	3.5	n.a.	4.5
Italy	0.1	—	—	0.1	0.1	0.4	—	1.0	0.7	8.7
UK	0.9	.	0.3	.	—	0.1	0.1	1.4	1.4	1.4
Other OECD Europe[3]	—	—	—	—	0.7	3.1	0.2	4.0	1.8	11.5
Japan	0.1	0.2	0.7	0.2	0.1	—	0.7	2.0	—	58.0
Canada	7.9	—	—	—	0.1	—	0.7	7.9	7.4	14.9
US	—	0.1	—	—	—	—	—	0.1	n.a.	0.1

1. Thermal coal is coal not delivered to coking plants.
2. Excludes intra-community trade.
3. Exludes some over the frontier imports of brown coal and lignite for electricity generation.
4. Country breakdowns refers only to public power stations. Total includes imports from third party countries by autoproducers.
5. Coal actually used by Ontario Hydro.
— nil.
. less than 0.05 Mt.

Annex 2 Table 2-3 **Thermal[1] Coal Imports of OECD Countries (1973-76)** (cont'd.)

b) 1974

Importing Area	USA	Canada	Australia	South Africa	USSR	Poland	Other	Total thermal coal imports	Delivered to electric sector	Total hard coal imports
EEC[2]	3.0	0.4	1.7	1.2	2.7	9.0	0.8	18.7	11.7[4]	38.0
of which:										
Denmark	—	0.1		—	0.3	3.0	0.1	3.5	2.7	3.5
France	0.1	0.1	0.4	0.4	1.5	1.9	0.1	4.5	2.6	8.8
Germany	1.9	0.1	·	0.4	0.2	1.9	0.2	4.7	n.a.	4.8
Italy	0.1	·	—	0.1	0.3	0.4	·	1.0	1.0	9.3
UK	0.8	·	1.0	—	—	0.7	0.1	2.6	2.6	3.5
Other OECD Europe[3]	—	—	—	·	0.7	0.4	0.4	5.3	2.0	
Japan	0.2	0.1	0.8	0.2	0.2	—	1.3	2.8	0.1	
Canada	5.6	—	—	—	—	—	—	5.6	5.3[6]	
US	—	0.5	0.1	0.2	—	0.6	1.9	1.9	n.a.	1.9

6. Coal actually used by Ontario Hydro was 7.3 Mt.

Annex 2 Table 2-3 **Thermal[1] Coal Imports of OECD Countries (1973-76)** (cont'd.)

c) 1975

Importing Area	Exporting Area							Total thermal coal imports	Delivered to electric sector	Total hard coal imports
	USA	Canada	Australia	South Africa	USSR	Poland	Other			
EEC[2]	5.8	0.8	3.1	1.6	2.5	9.0	0.6	23.4	17.8[4]	41.1
of which:										
Denmark	0.1	0.1	—	—	0.4	3.5	—	4.1	3.5	4.1
France	1.3	0.3	0.3	0.7	0.6	2.2	0.1	6.8	4.5	10.9
Germany	2.6	0.1	0.2	0.5	0.1	1.9	0.2	5.6	3.8	5.8
Italy	0.4	—	—	0.1	0.2	0.6	.	1.3	1.0	9.6
UK	1.4	0.3	2.3	.	—	—	0.2	4.2	4.1	5.0
Other OECD										
Europe[3]	—	—	—	—	0.6	4.0	0.1	4.6	2.9	13.5
Japan	0.1	0.2	1.0	—	0.1	0.1	1.0	2.5	0.2	62.3
Canada	8.9	—	—	—	—	—	—	8.9	8.5[7]	15.4
US	—	0.2	0.1	0.3	—	0.2	—	0.9	0.9	0.9

7. See note 6. Coal used was 7.3 Mt.

Annex 2 Table 2-3 **Thermal[1] Coal Imports of OECD Countries (1973-76)** (cont'd.)

d) 1976

Importing Area	USA	Canada	Australia	South Africa	USSR	Poland	Other	Total thermal coal imports	Delivered to electric sector	Total hard coal imports
EEC[2]	3.6	0.6	1.6	3.5	3.0	11.3	0.2	23.8	18.8[4]	43.7
of which:										
Denmark	—	0.2	—	—	.	0.6	3.2	4.2	3.4	4.2
France	1.1	0.1	0.4	2.0	.	1.5	4.1	9.2	6.9	13.7
Germany	1.7	0.3	.	0.7	.	0.2	2.1	5.3	3.6	5.4
Italy	—	—	—	0.4	.	0.2	0.9	1.5	1.2	10.0
UK	—	—	1.1	.	.	—	.	1.3	1.3	2.4
Other OECD Europe[3]	—	—	—	0.1	0.8	2.8	0.1	3.8	2.9	13.7
Japan	0.1	0.2	0.8	—	0.4	—	0.9	2.4	0.2	60.1
Canada	8.2	.	—	—	—	—	—	8.2	7.8	14.6
US	—	.	—	0.7	—	0.3	.	1.1	1.1	1.1

LIST OF SYMBOLS AND ABBREVIATIONS

AFB	= Atmospheric Fluidized Combustors	MHD	= Magnetohydrodynamic
BACT	= Best Available Control Technology	MFB	= Major Fuel Burning Industrial Plants
b/d	= barrels per day	Mt	= Million tons
boe/d	= barrels of oil equivalent per day	Mtce	= Million tons of coal equivalent
BTU	= British Thermal Unit	Mtoe	= Million tons of oil equivalent
Cal	= Calorie	NCB	= National Coal Board (UK)
CIF	= Cost including insurance and freight	NGL	= Natural Gas Liquids
CPE	= Centrally Planned Economies	NM	= Nautical Mile
DOE	= Department of Energy (U.S.)	NSPS	= New Source Performance Standard
dwt	= Deadweight Tons	OBO	= Ore/Bulk/Oil Carriers
Escom	= Electricity Supply Commission (South Africa)	OTA	= Office of Technology Assessment (U.S.)
FBC	= Fluidized Bed Combustion	p.a.	= per annum
FGD	= Flue Gas Desulphurization	SNG	= Synthetic Natural Gas
FOB	= Free on Board	T	= Tera = trillion = million million = 10^{12}
G	= Giga = billion = thousand million = 10^9	t	= ton (metric)
GDP	= Gross Domestic Product	TER	= Total Energy Requirements
GW	= Gigawatts	tce	= ton of coal equivalent
kW	= Kilowatt	toe	= ton of oil equivalent
kWh	= Kilowatt-hour	USEPA	= U.S. Environmental Protection Agency
LDC	= Less Developed Countries		
M	= Mega = million	W	= Watt
MESA	= Mining Enforcement and Safety Administration (U.S.)	WEC	= World Energy Conference

Approximate Conversion Factors

To convert : / Into :	Mtoe	Mtce	Mb/d	GW
	Multiplied by :			
Mtoe = 10^{13} kcal	1	1.43	0.02	0.72
Mtce = 7×10^{12} kcal	0.7	1.	0.01	0.5
Mb/d	50.0	71.4	1	35.7
GW	1.4	2.0	0.03	1

Note : These conversion factors apply to annual operation at an OECD-average capacity utilization of 65% and a conversion efficiency of 35%: actual capacity utilization and conversion efficiencies will, however, vary from country to country and over time. Furthermore, the nuclear capacity (GW) figures are usually expressed as installed capacity at year-end, while the other terms are based on year long operation. Therefore, converting these other terms to nuclear capacity will usually understate slightly year-end installed capacity.

Requests for further information regarding this study can be addressed to : Energy Economic Analysis Division, IEA-OECD; 2, rue André-Pascal; 75775 PARIS CEDEX 16.

OECD PUBLICATIONS, 2, rue André-Pascal, 75775 Paris Cedex 16 - No. 41015 1978
PRINTED IN FRANCE